At Sylvan, we believe that a lifelong love of learning begins at an early age, and we are glad you have chosen our resources to help your children experience the joy of mathematics as they build critical reasoning skills. We know that the time you spend with your children reinforcing the lessons learned in school will contribute to their love of learning.

Success in math requires more than just memorizing basic facts and algorithms; it also requires children to make sense of size, shape, and numbers as they appear in the world. Children who can connect their understanding of math to the world around them will be ready for the challenges of mathematics as they advance to more complex topics.

We use a research-based, step-by-step process in teaching math at Sylvan that includes thought-provoking math problems and activities. As students increase their success as problem solvers, they become more confident. With increasing confidence, students build even more success. The design of the Sylvan workbooks will help you to help your children build the skills and confidence that will contribute to success in school.

Included with your purchase of this workbook is a coupon for a discount at a participating Sylvan center. We hope you will use this coupon to further your children's academic journeys. Let us partner with you to support the development of confident, well-prepared, independent learners.

The Sylvan Team

Sylvan Learning Center
Unleash your child's potential here

No matter how big or small the academic challenge, every child has the ability to learn. But sometimes children need help making it happen. Sylvan believes every child has the potential to do great things. And, we know better than anyone else how to tap into that academic potential so that a child's future really is full of possibilities. Sylvan Learning Center is the place where your child can build and master the learning skills needed to succeed and unlock the potential you know is there.

The proven, personalized approach of our in-center programs deliver unparalleled results that other supplemental education services simply can't match. Your child's achievements will be seen not only in test scores and report cards but outside the classroom as well. And when he starts achieving his full potential, everyone will know it. You will see a new level of confidence come through in everything he does and every interaction he has.

How can Sylvan's personalized in-center approach help your child unleash his potential?

- Starting with our exclusive Sylvan Skills Assessment®, we pinpoint your child's exact academic needs.
- Then we develop a customized learning plan designed to achieve your child's academic goals.
- Through our method of skill mastery, your child will not only learn and master every skill in his personalized plan, he will be truly motivated and inspired to achieve his full potential.

To get started, included with this Sylvan product purchase is $10 off our exclusive Sylvan Skills Assessment®. Simply use this coupon and contact your local Sylvan Learning Center to set up your appointment.

And to learn more about Sylvan and our innovative in-center programs, call 1-800-EDUCATE or visit www.SylvanLearning.com. ***With over 1,000 locations in North America, there is a Sylvan Learning Center near you!***

2nd Grade Math Games & Puzzles

Published in the United States by Random House, Inc., New York, and in Canada by Random House of Canada Limited, Toronto.

www.tutoring.sylvanlearning.com

Created by Smarterville Productions LLC
Producer & Editorial Direction: The Linguistic Edge
Producer: TJ Trochlil McGreevy
Writer: Amy Kraft
Cover and Interior Illustrations: Shawn Finley and Duendes del Sur
Layout and Art Direction: SunDried Penguin
Director of Product Development: Russell Ginns

First Edition

ISBN: 978-0-375-43037-4

This book is available at special discounts for bulk purchases for sales promotions or premiums. For more information, write to Special Markets/Premium Sales, 1745 Broadway, MD 6-2, New York, New York 10019 or e-mail specialmarkets@randomhouse.com.

PRINTED IN CHINA

10 9 8 7 6 5 4 3 2 1

Contents

Hidden Design

COUNT the hundreds, tens, and ones. Then COLOR the squares that match the numbers to see the hidden design.

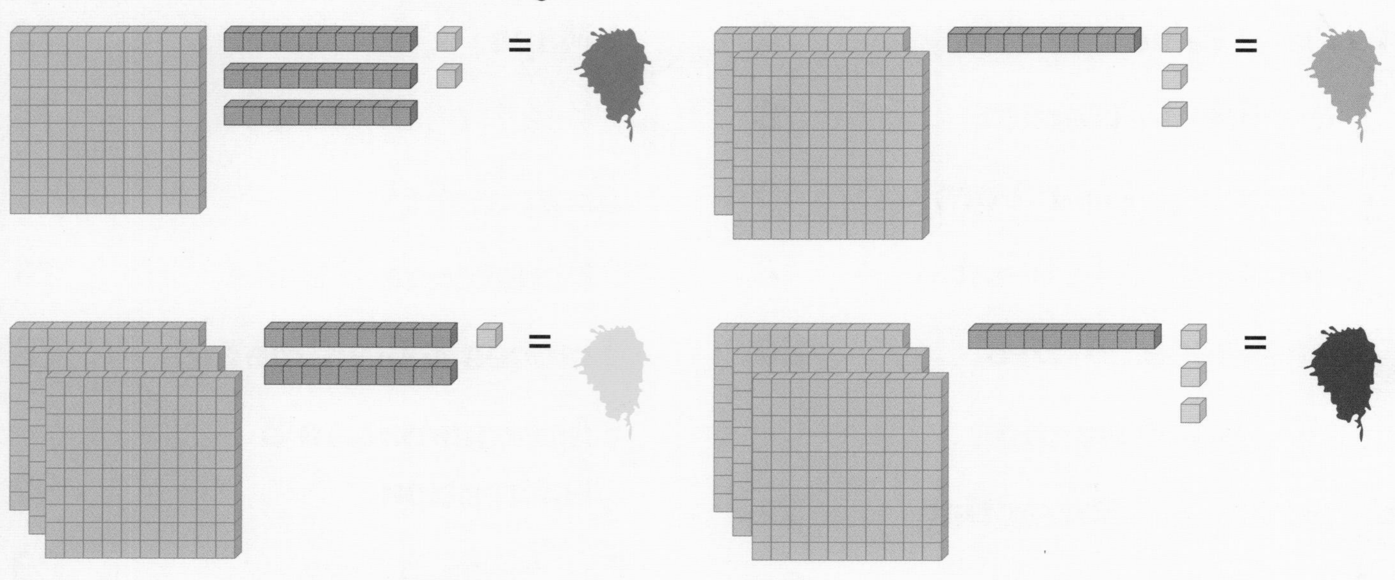

313	213	132	321	132	213	313	321
213	132	321	313	321	132	213	313
132	321	313	213	313	321	132	213
321	313	213	132	213	313	321	132
313	213	132	321	132	213	313	321
213	132	321	313	321	132	213	313
132	321	313	213	313	321	132	213
321	313	213	132	213	313	321	132

Safe Crackers

WRITE the number for each picture. Then WRITE the digit from the hundreds place of each number from left to right to find the combination for the safe.

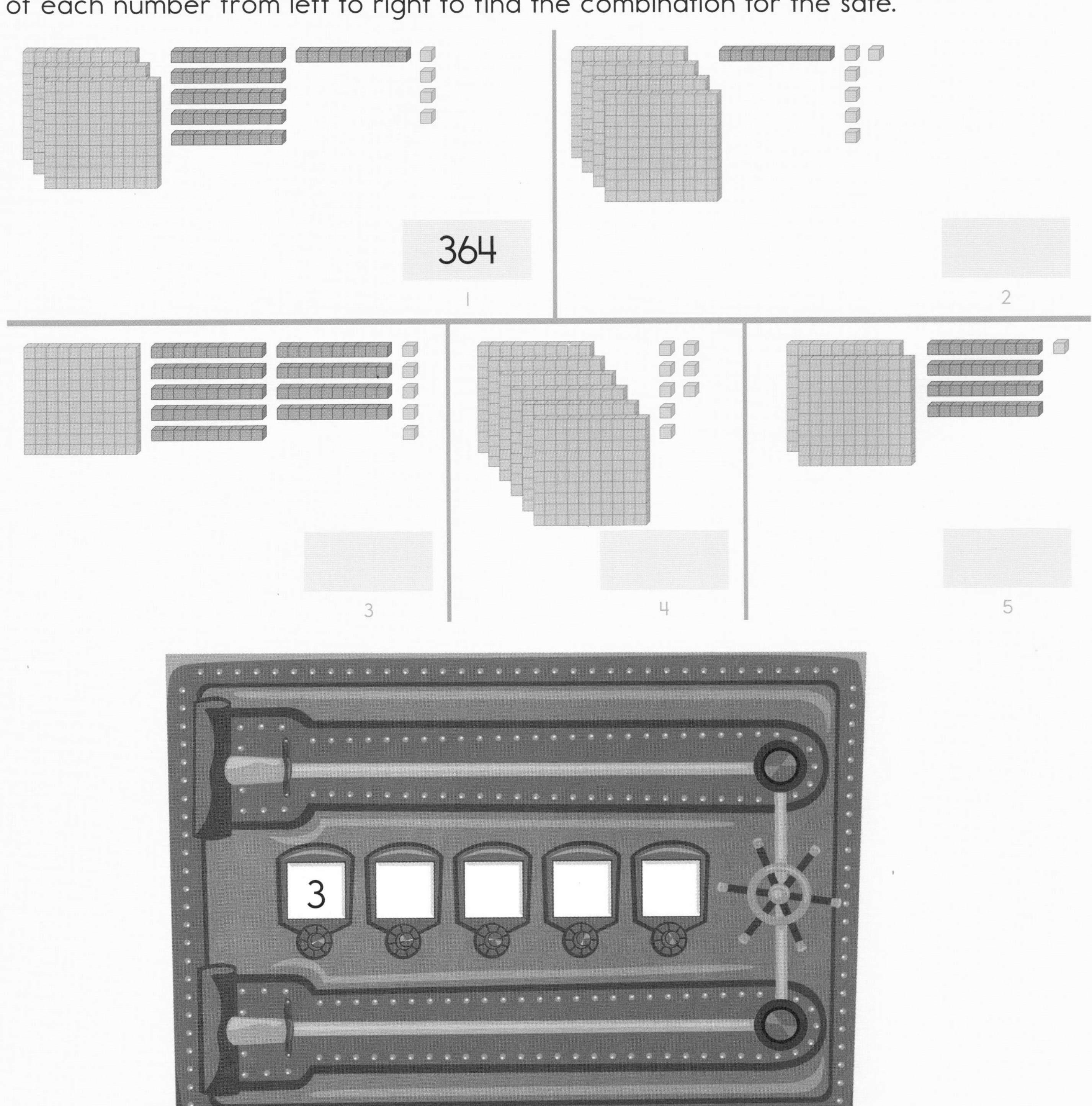

Number Search

WRITE the number for each picture. Then CIRCLE it in the puzzle.

HINT: Numbers are across and down only.

159

1

2

3

4

5

6

5	8	0	3	1	4
2	1	5	4	7	0
0	5	0	6	8	3
8	9	3	1	3	5
3	3	7	6	0	0
1	5	2	3	4	8
7	1	8	4	2	1
1	9	6	4	7	2

Roll It

ROLL a number cube, and WRITE the number in the first box. ROLL the number cube two more times, and WRITE the numbers in the second and third boxes. Then COLOR the hundreds, tens, and ones to match the number.

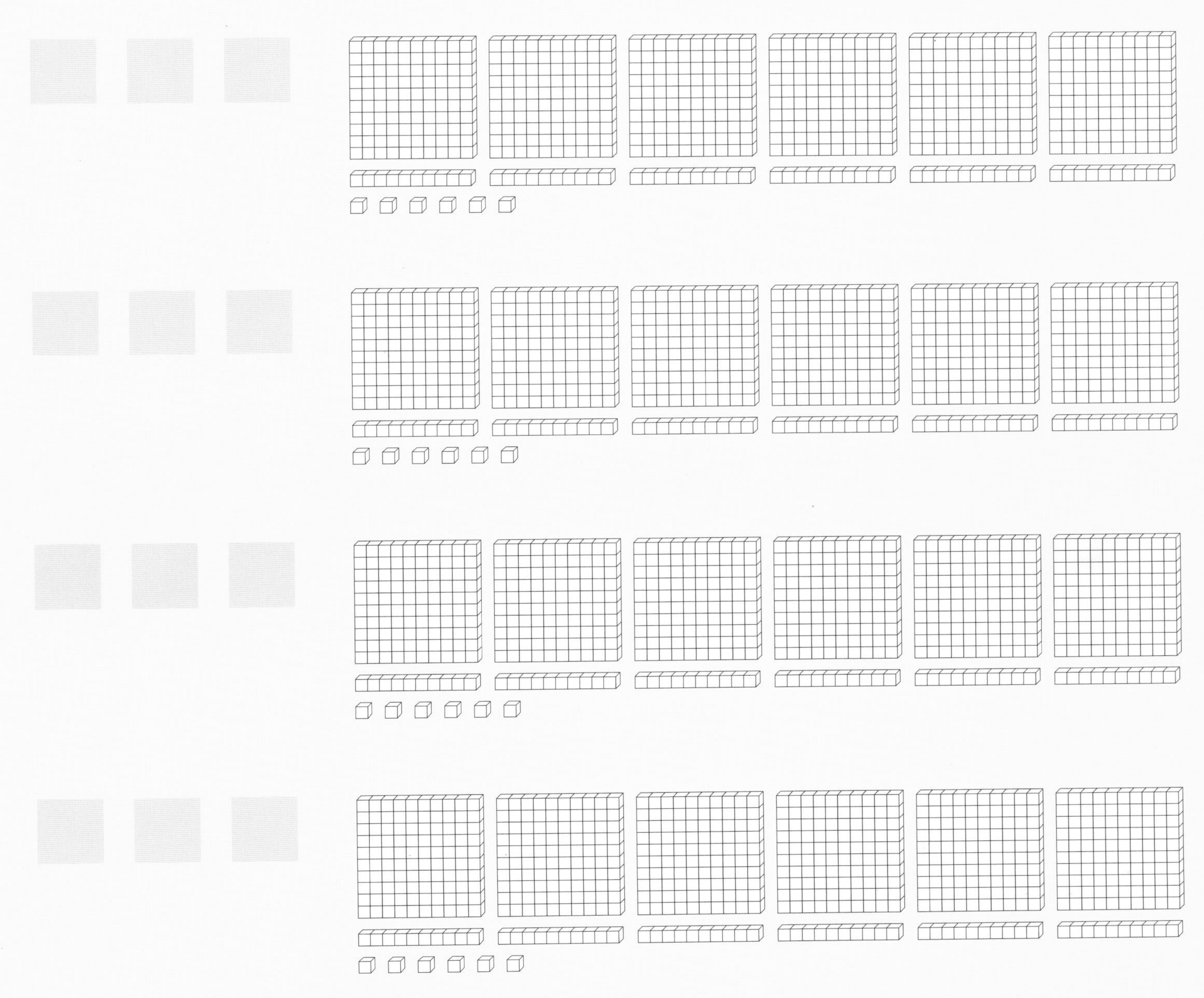

Code Breaker

WRITE the missing numbers. Then WRITE the letter that matches each number to solve the riddle.

1.

97	98	99		101	102		104
B			Y			K	

2.

213	214		216	217		219	
		I			R		U

3.

		748	749	750	751		
T	O					L	E

What did the zero say to the eight?

___ 215 ___ 752 ___ 215 ___ 103 ___ 753

___ 100 ___ 747 ___ 220 ___ 218   B 97 ___ 753 ___ 752 ___ 746 !

Skipping Stones

DRAW a path by skip counting by 5 to cross the river.

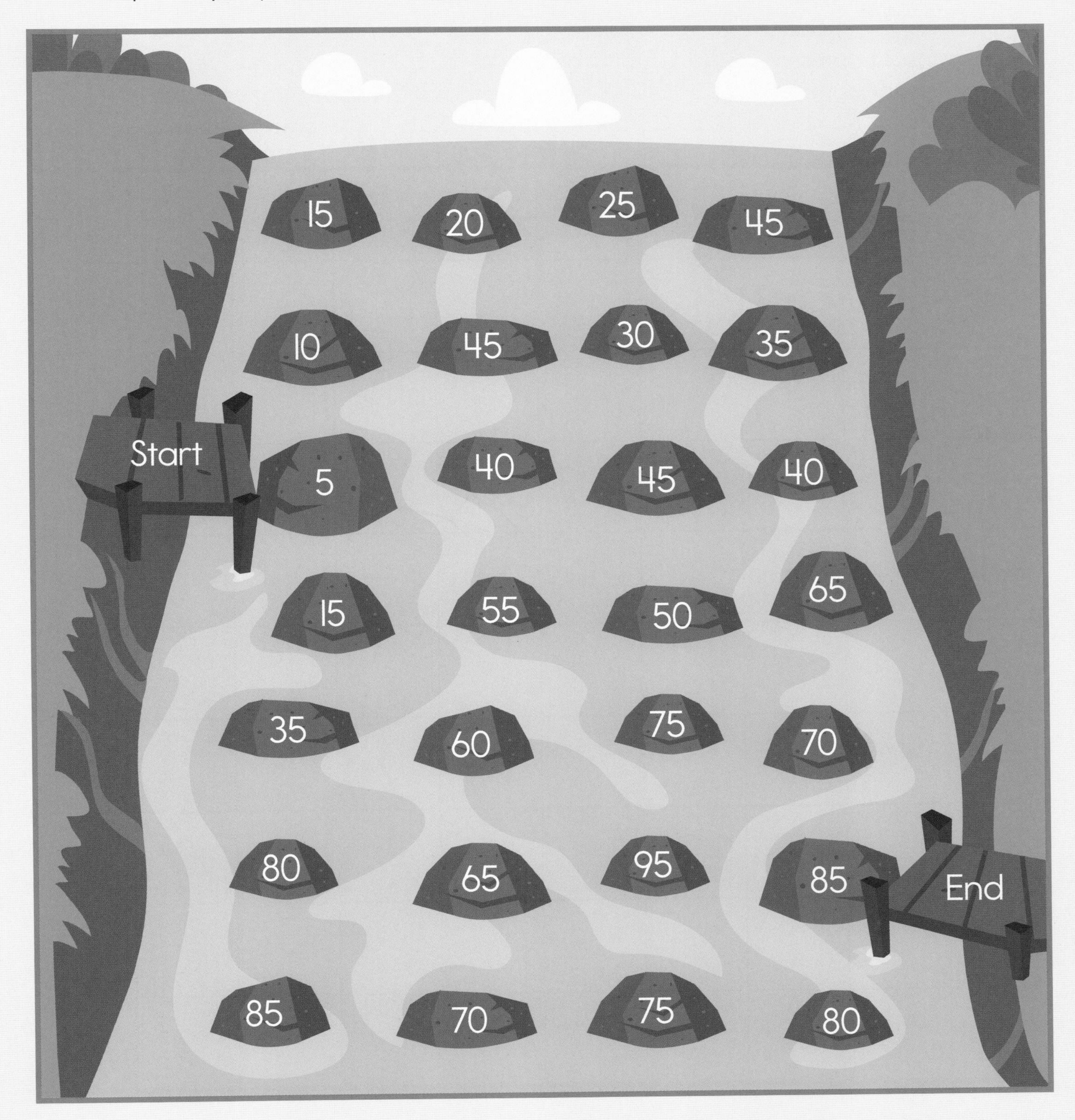

Three for Thrills

WRITE the numbers in the hundreds chart. Then COLOR the chart by following the directions.

1. Starting at number 1, SKIP COUNT by 3 and COLOR the squares yellow.
2. Starting at number 2, SKIP COUNT by 3 and COLOR the squares blue.
3. Starting at number 3, SKIP COUNT by 3 and COLOR the squares orange.

1	2								10
11									20
21									30
31									40
41									50
51									60
61									70
71									80
81									90
91									100

Skip to My Loo

SKIP COUNT by 10, 4, and 7, and WRITE the numbers along each track.

Skip count by:

10	4	7
10	4	7
20		

Finish

Just Right

WRITE each number next to a smaller blue number.

HINT: There may be more than one place to put a number, but you need to use every number.

742 113 981 187 256 677 409 556 823 399

599

1

98

2

724

3

545

4

750

5

251

6

830

7

400

8

178

9

398

10

Totally Tangled

Each numbered circle is connected to another numbered circle. FIND the pairs of numbers, and COLOR the circle with the smaller number.

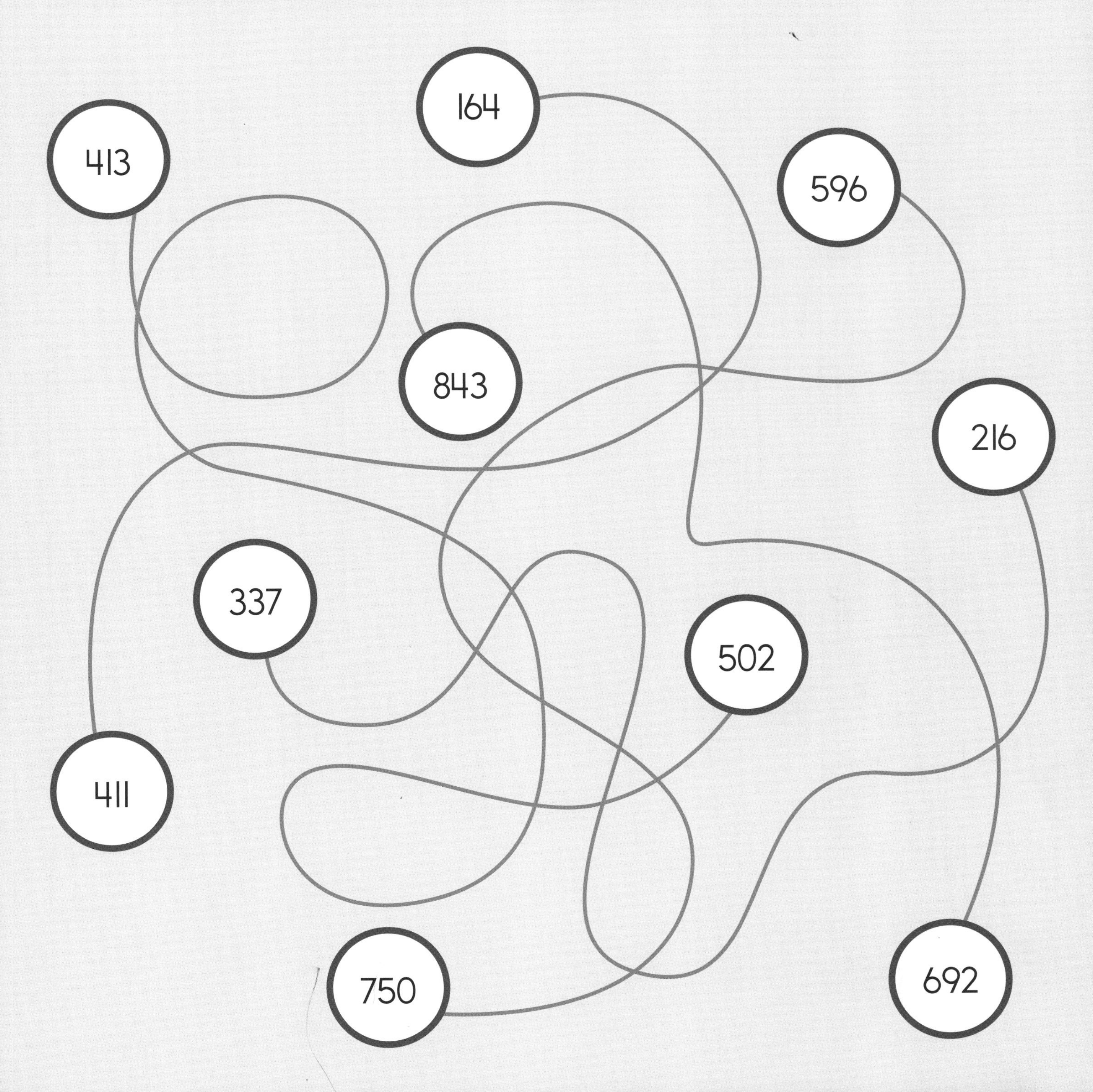

Win Big

Wherever two boxes point to one box, WRITE the larger number. Start at the sides and work toward the center to see which number will win big.

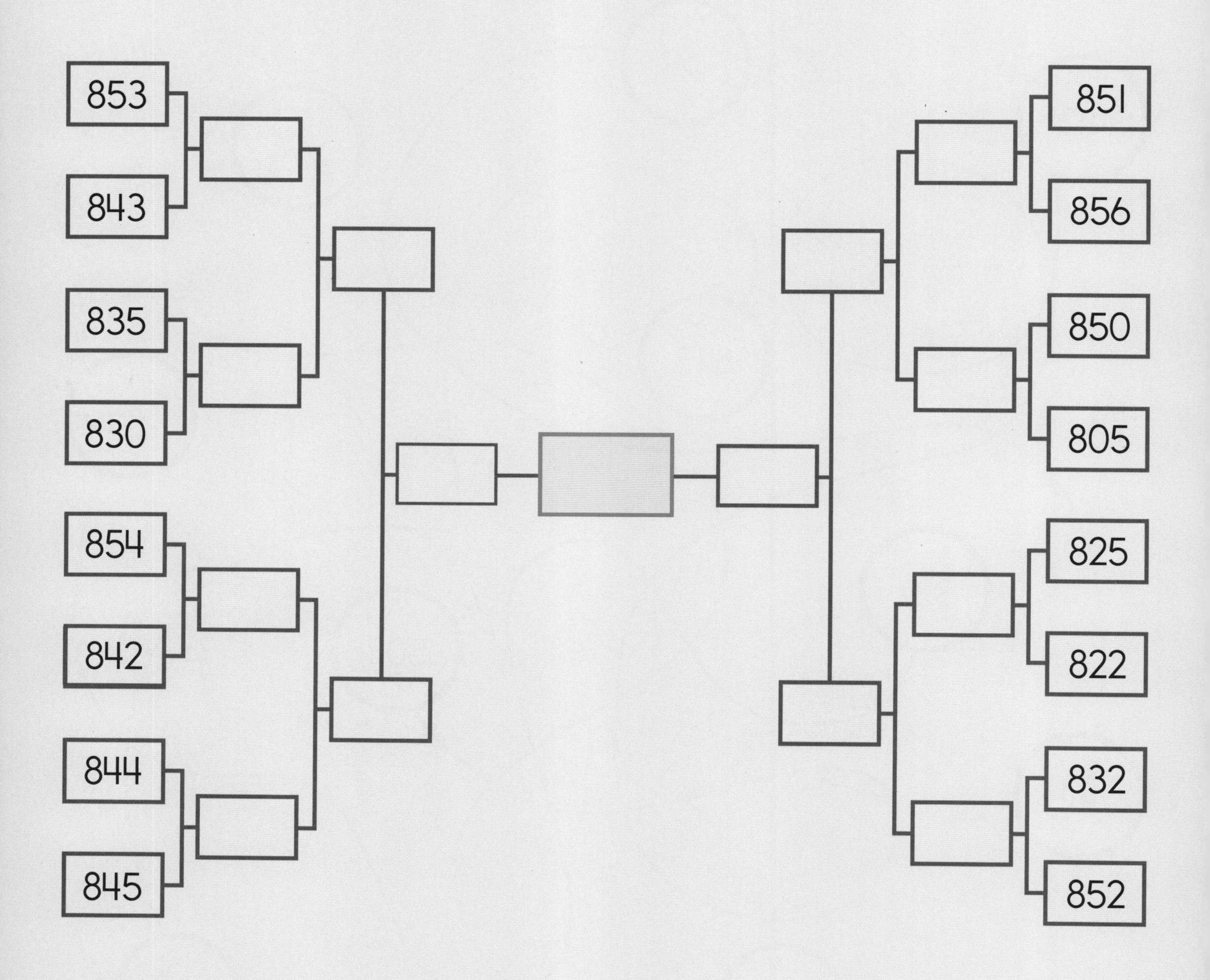

Totally Tangled

Each numbered circle is connected to another numbered circle. FIND the pairs of numbers, and COLOR the circle with the larger number.

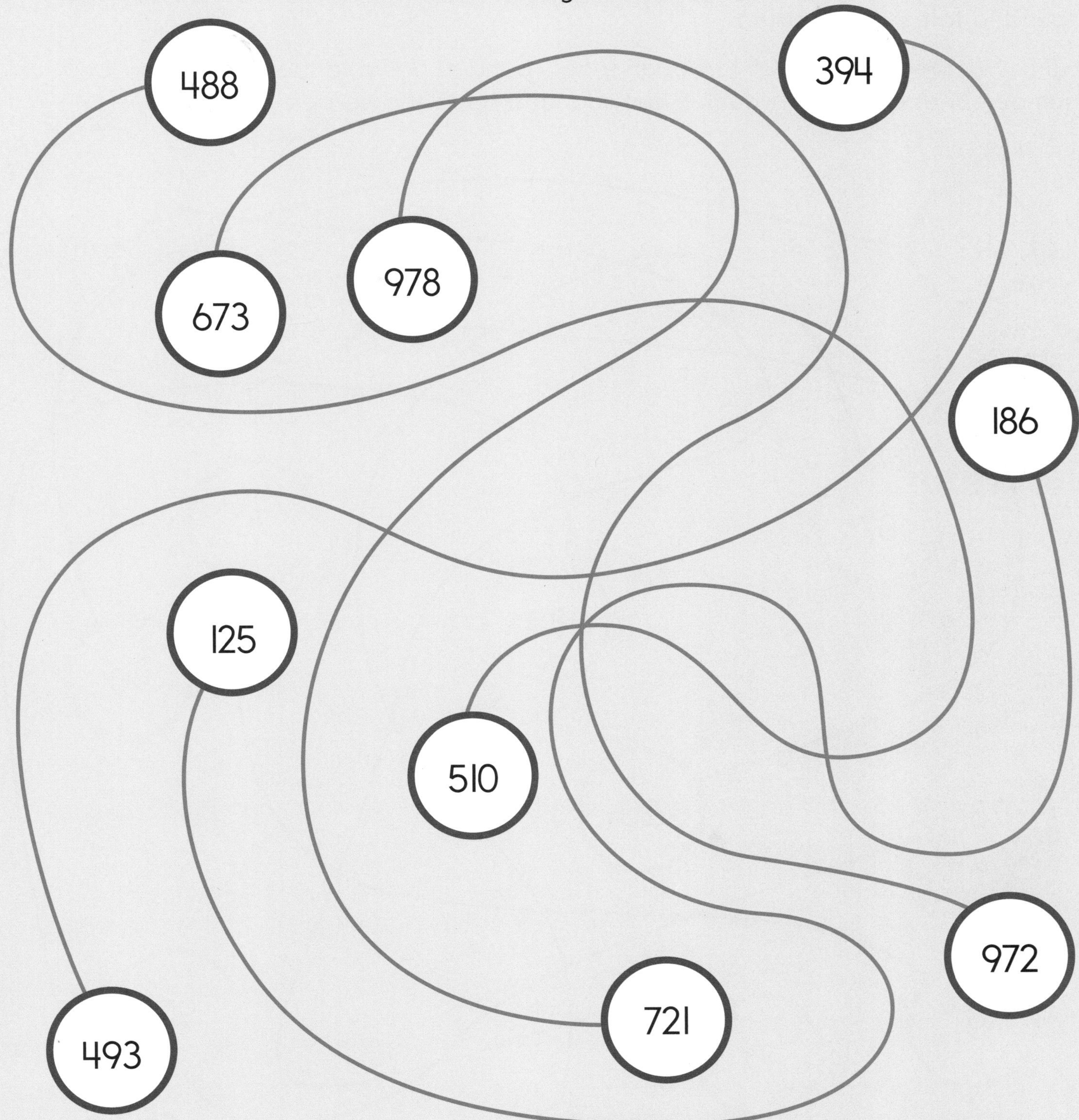

Totally Tangled

Each numbered circle is connected to another numbered circle. FIND the pairs of numbers, and COLOR any pair that shows a number with that number correctly rounded to the nearest ten.

HINT: Numbers that end in 1 through 4 get rounded down to the nearest ten, and numbers that end in 5 through 9 get rounded up to the nearest ten.

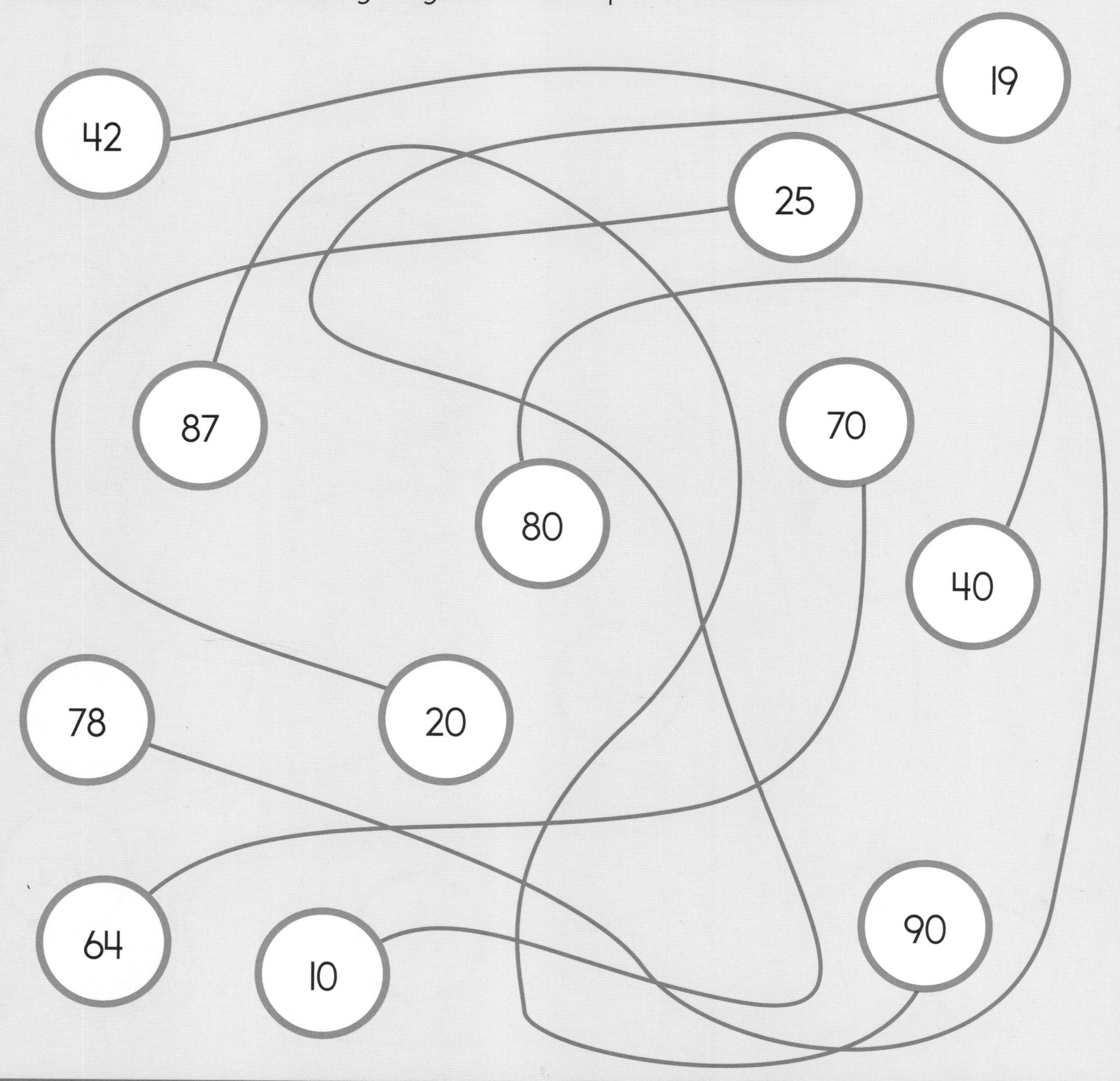

Roll It

ROLL a number cube, and WRITE the number in the first box. ROLL the number cube two more times, and WRITE the numbers in the second and third boxes. Then ROUND each number to the nearest ten and the nearest hundred.

HINT: Numbers that end in 1 through 49 get rounded down to the nearest hundred, and numbers that end in 50 through 99 get rounded up to the nearest hundred.

			Nearest Ten	Nearest Hundred

Just Right

WRITE each of the numbers to correctly complete the sentences. There may be more than one place to put a number, but you need to use every number.

HINT: Numbers that end in 1 through 49 get rounded down to the nearest hundred, and numbers that end in 50 through 99 get rounded up to the nearest hundred.

549	278	709	751	952
717	932	285	544	

1. ________ rounded to the nearest hundred is 300.
2. ________ rounded to the nearest ten is 540.
3. ________ rounded to the nearest hundred is 700.
4. ________ rounded to the nearest ten is 950.
5. ________ rounded to the nearest hundred is 500.
6. ________ rounded to the nearest ten is 280.
7. ________ rounded to the nearest hundred is 900.
8. ________ rounded to the nearest ten is 720.
9. ________ rounded to the nearest hundred is 800.

Fitting In

GUESS how many marbles are needed to fill the circle. WRITE your guess. Then turn the page to CHECK your guess.

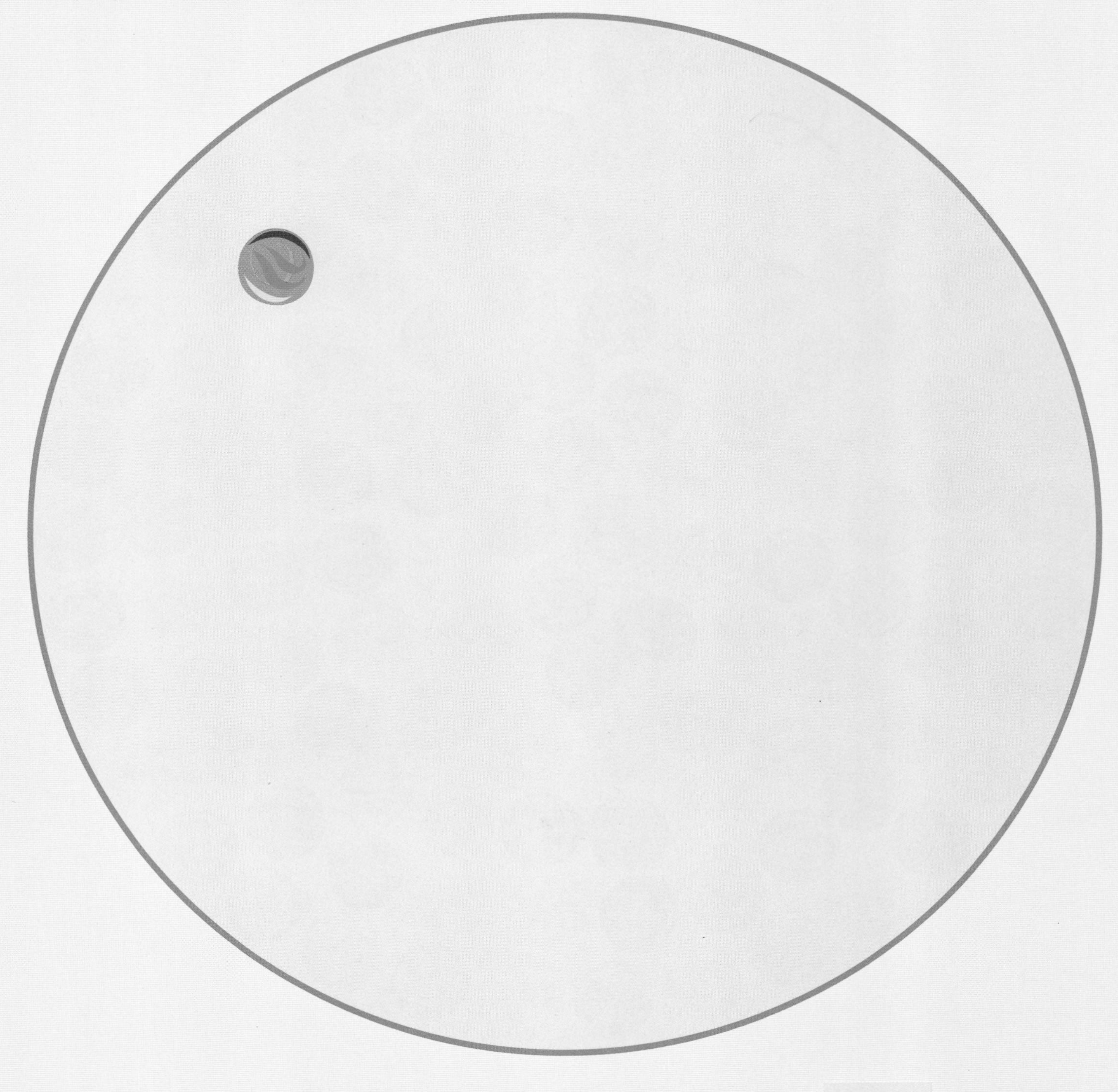

Guess: ______ marbles

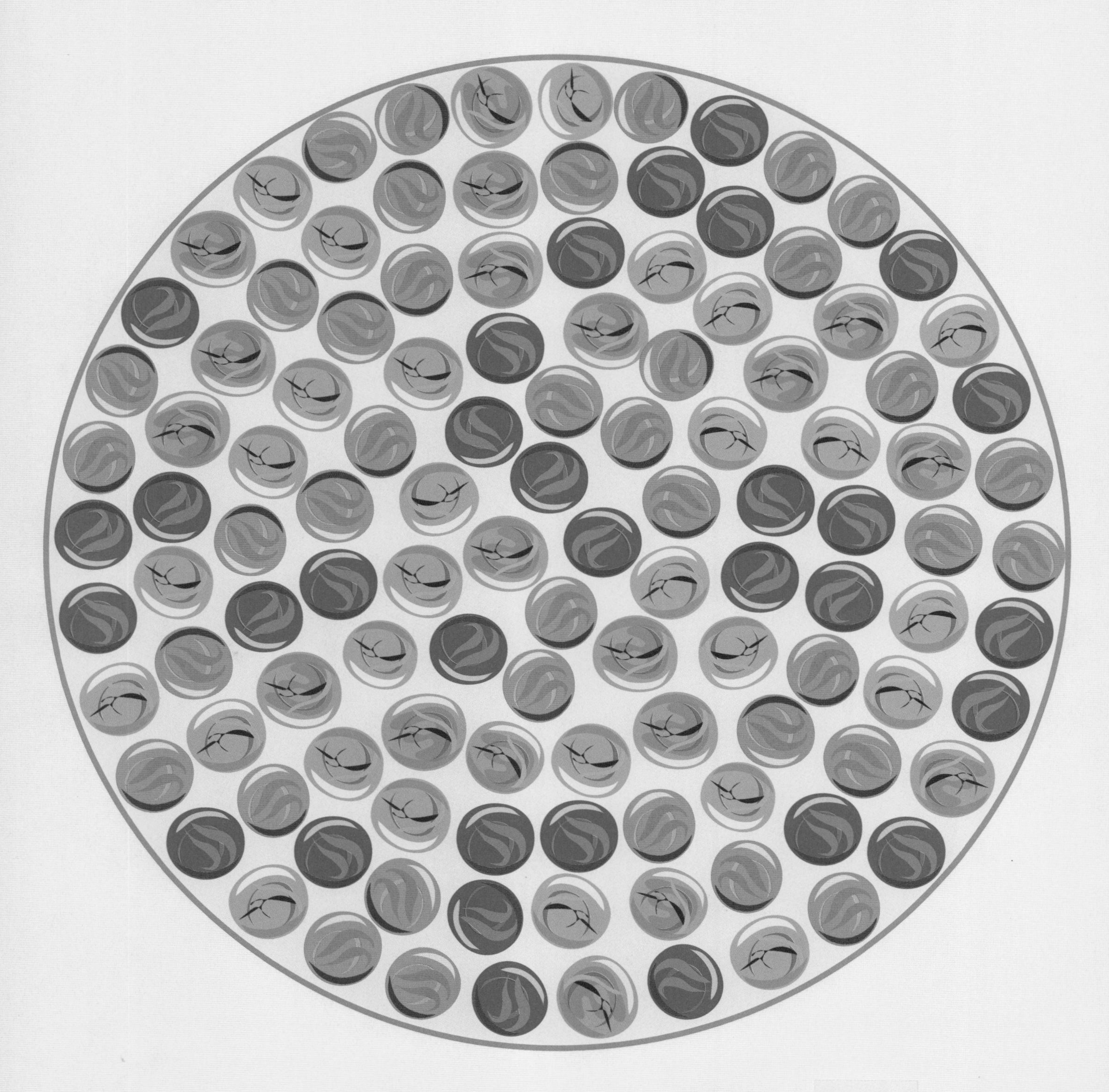

Check: marbles

Fitting In

GUESS how many pennies are needed to fill the square. WRITE your guess. Then fill the square with pennies to CHECK your guess.

Guess: ______ pennies

Check: ______ pennies

Spiraling Sequence

SKIP COUNT by 6, and WRITE the numbers. Can you finish the spiral to the center?

6	12	18				

Who Am I?

READ the clues, and CIRCLE the mystery number.

HINT: Cross out any number that does not match the clues.

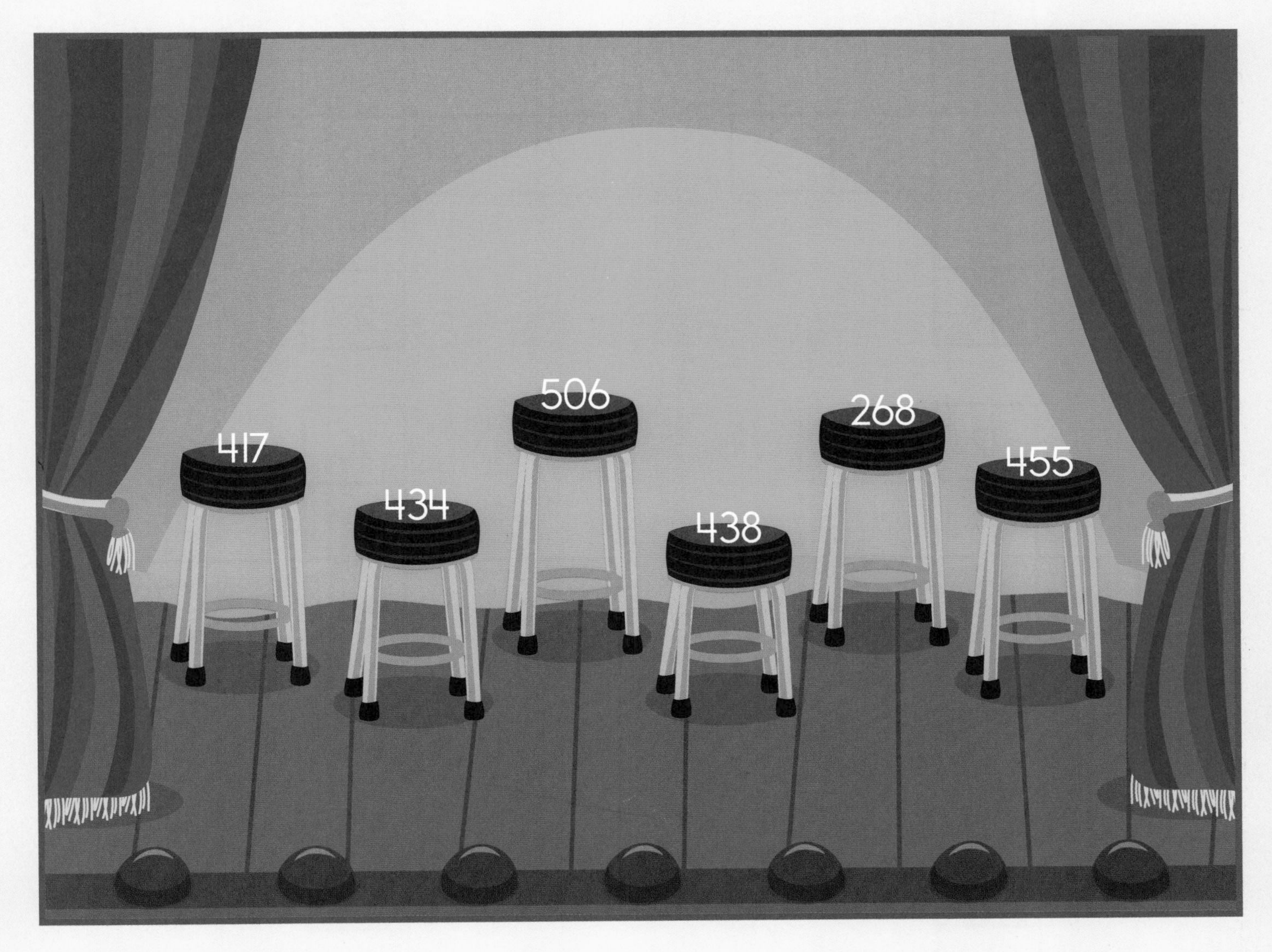

I am more than 300.

I am less than 500.

When rounded to the nearest hundred, I'm 400.

I have a 3 in the tens place.

When rounded to the nearest ten, I'm 440.

Who am I?

Missing Middles

WRITE the number missing from the center square.

1.

		32		
		+		
17	+		=	38
		=		
		53		

2.

		43		
		+		
50	+		=	66
		=		
		59		

3.

		12		
		+		
44	+		=	89
		=		
		57		

4.

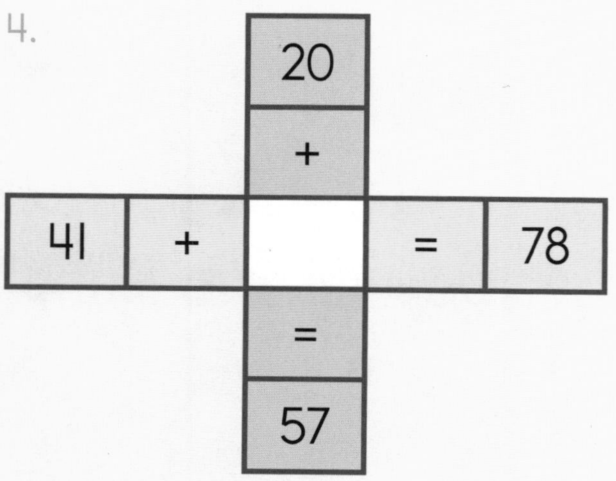

5.

		31		
		+		
23	+		=	85
		=		
		93		

6.

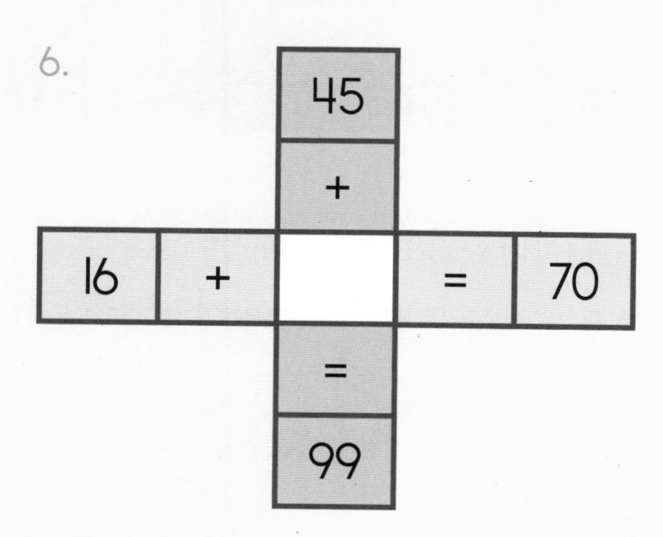

Crossing Paths

WRITE the missing numbers.

3 +	11	+ 5
= ◯ +	23	+ ◯ =
= ◯ +	20	+ ◯ =
= ◯ +	18	+ ◯ =
= ◯ +	14	+ ◯ =
= ◯		◯ =

4 +	37	+ 2
= ◯ +	12	+ ◯ =
= ◯ +	8	+ ◯ =
= ◯ +	25	+ ◯ =
= ◯ +	14	+ ◯ =
= ◯		◯ =

Super Square

WRITE numbers in the empty squares to finish all of the addition problems.

3	+	12	=	
+		+		+
9	+		=	
=		=		=
	+	37	=	

Code Breaker

SOLVE each problem. WRITE the letter that matches each sum to solve the riddle.

1. $11 + 26$ = ___ O	2. $38 + 30$ = ___ L	3. $46 + 13$ = ___ V	4. $21 + 58$ = ___ C
5. $43 + 52$ = ___ P	6. $17 + 31$ = ___ I	7. $81 + 12$ = ___ N	8. $67 + 10$ = ___ A
9. $29 + 34$ = ___ T	10. $13 + 58$ = ___ G	11. $62 + 18$ = ___ E	12. $77 + 14$ = ___ S

Where does the pencil go on vacation?

___ ___   ___ ___ ___ ___   ___ ___
48 63   71 37 80 91   63 37

___ ___ ___ ___ ___ ___ ___ ___ ___ ___ ___.
95 80 93 79 48 68 59 77 93 48 77

Pipe Down

WRITE each number. Then FOLLOW the pipe, and WRITE the same number in the next problem.

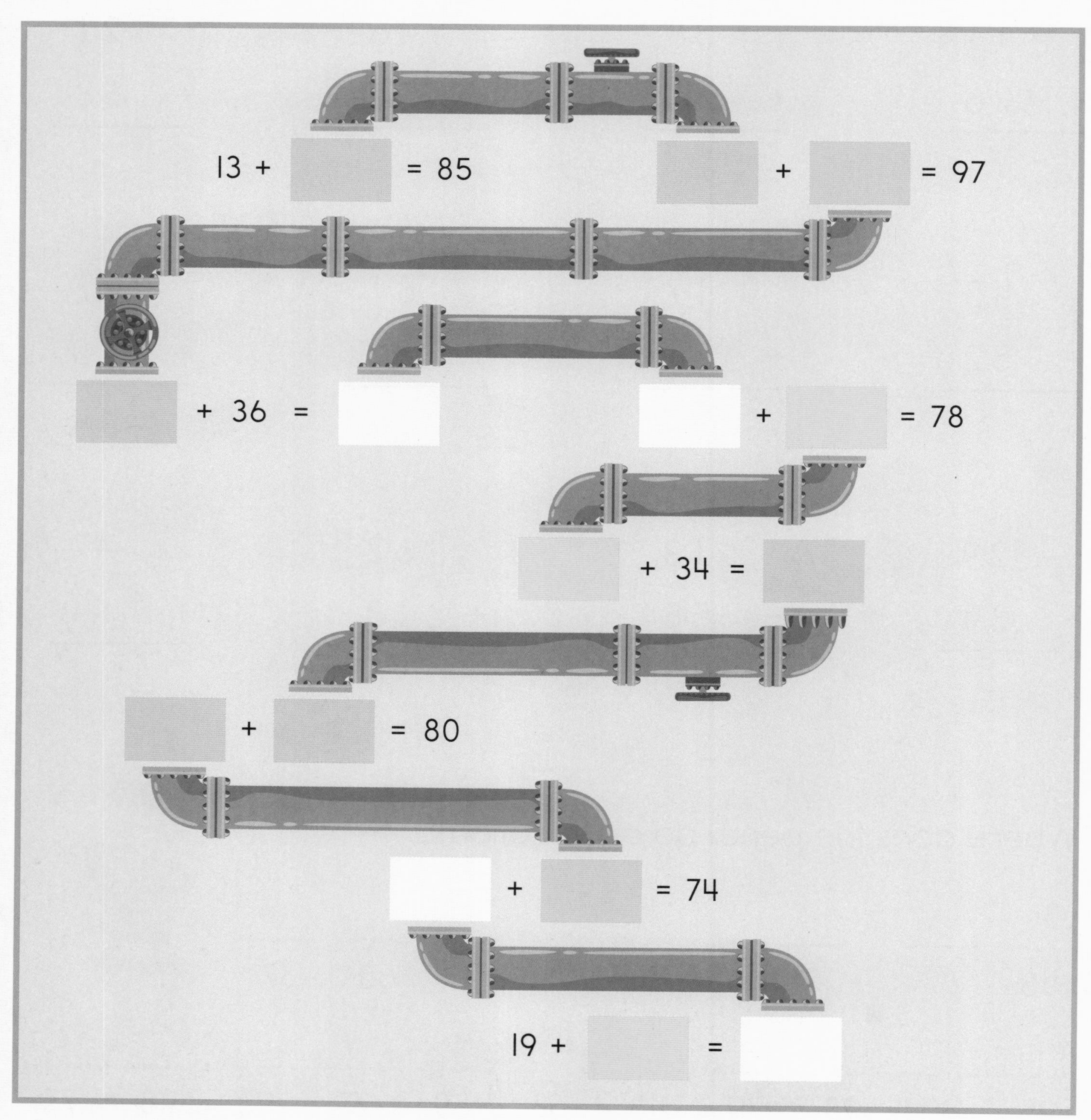

Ground Floor

DRAW a line connecting one of the numbers in the windows at the top with the sum on the ground floor. CHOOSE one operation from each floor on the way down.

HINT: When you're done, can you find more ways to the ground floor?

Missing Middles

WRITE the number missing from the center square.

1.

48
–
25 –   = 13
=
36

2.

37
–
78 –   = 45
=
4

3.

86
–
99 –   = 28
=
15

4.

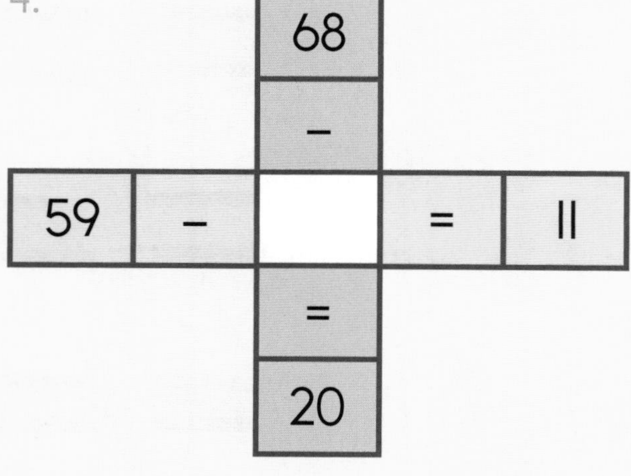

5.

95
–
44 –   = 22
=
73

6.

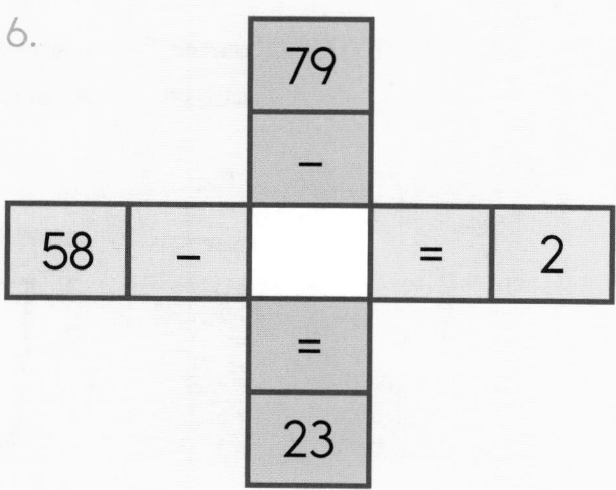

Crossing Paths

WRITE the missing numbers.

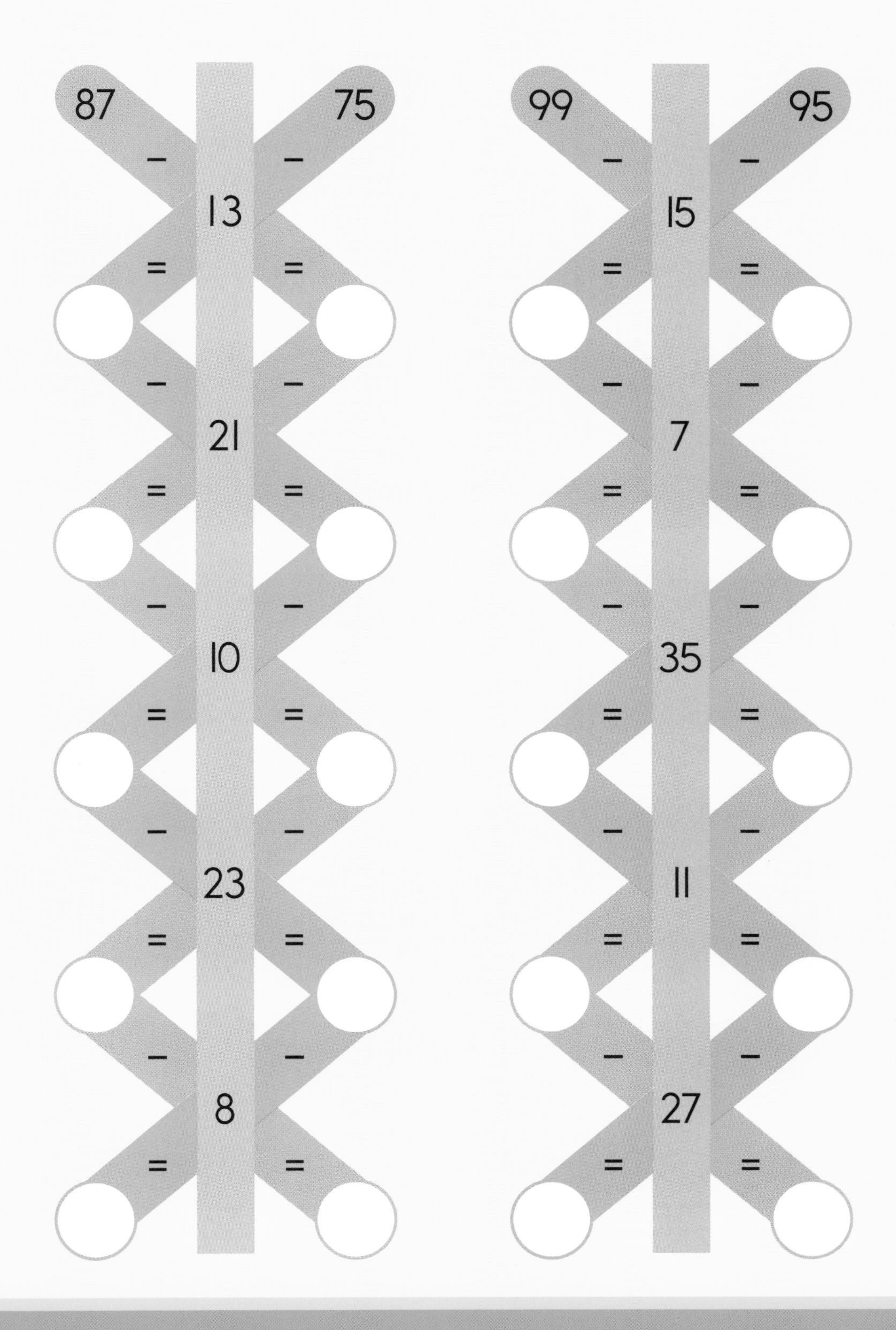

Super Square

WRITE numbers in the empty squares to finish all of the subtraction problems.

95	–		=	32
–		–		–
47	–		=	
=		=		=
	–	46	=	

Code Breaker

SOLVE each problem. WRITE the letter that matches each difference to solve the riddle.

1	2	3	4	5	6
42 − 11 = ___	75 − 52 = ___	18 − 16 = ___	84 − 62 = ___	91 − 45 = ___	72 − 23 = ___
W	D	O	A	L	V

7	8	9	10	11	12
63 − 37 = ___	71 − 19 = ___	72 − 59 = ___	96 − 49 = ___	60 − 24 = ___	52 − 36 = ___
U	T	S	Y	E	H

If you took two toys away from seven toys, how many toys would you have?

___ ___ ___   ___ ___ ___ ___ ___
47 2 26   31 2 26 46 23

___ ___ ___ ___   ___ ___ ___
16 22 49 36   52 31 2

___ ___ ___ ___.
52 2 47 13

Pipe Down

WRITE each number. Then FOLLOW the pipe, and WRITE the same number in the next problem.

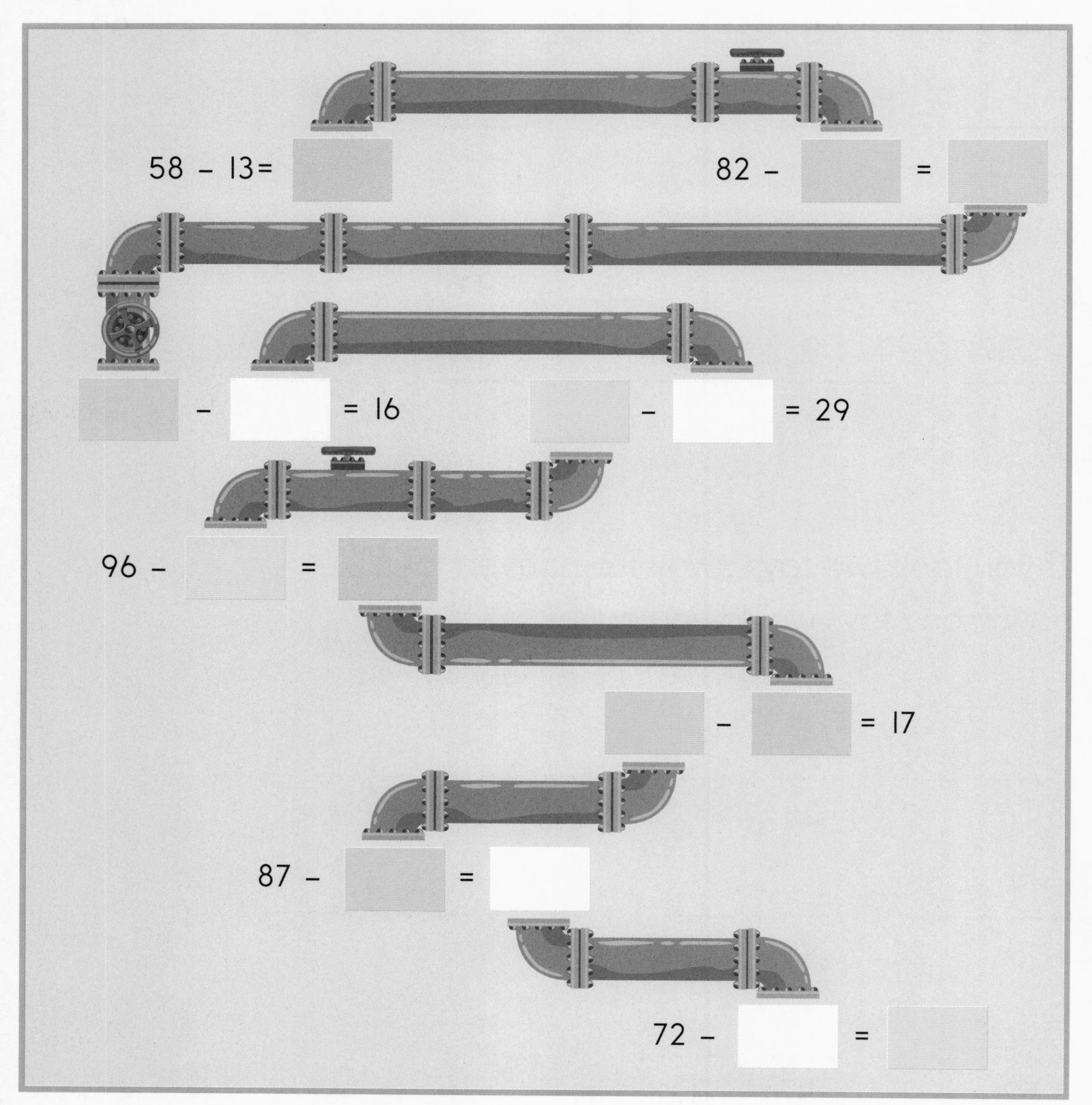

Ground Floor

DRAW a line connecting one of the numbers in the windows at the top with the difference on the ground floor. CHOOSE one operation from each floor on the way down.

HINT: When you're done, can you find more ways to the ground floor?

Crossing Paths

WRITE the missing numbers.

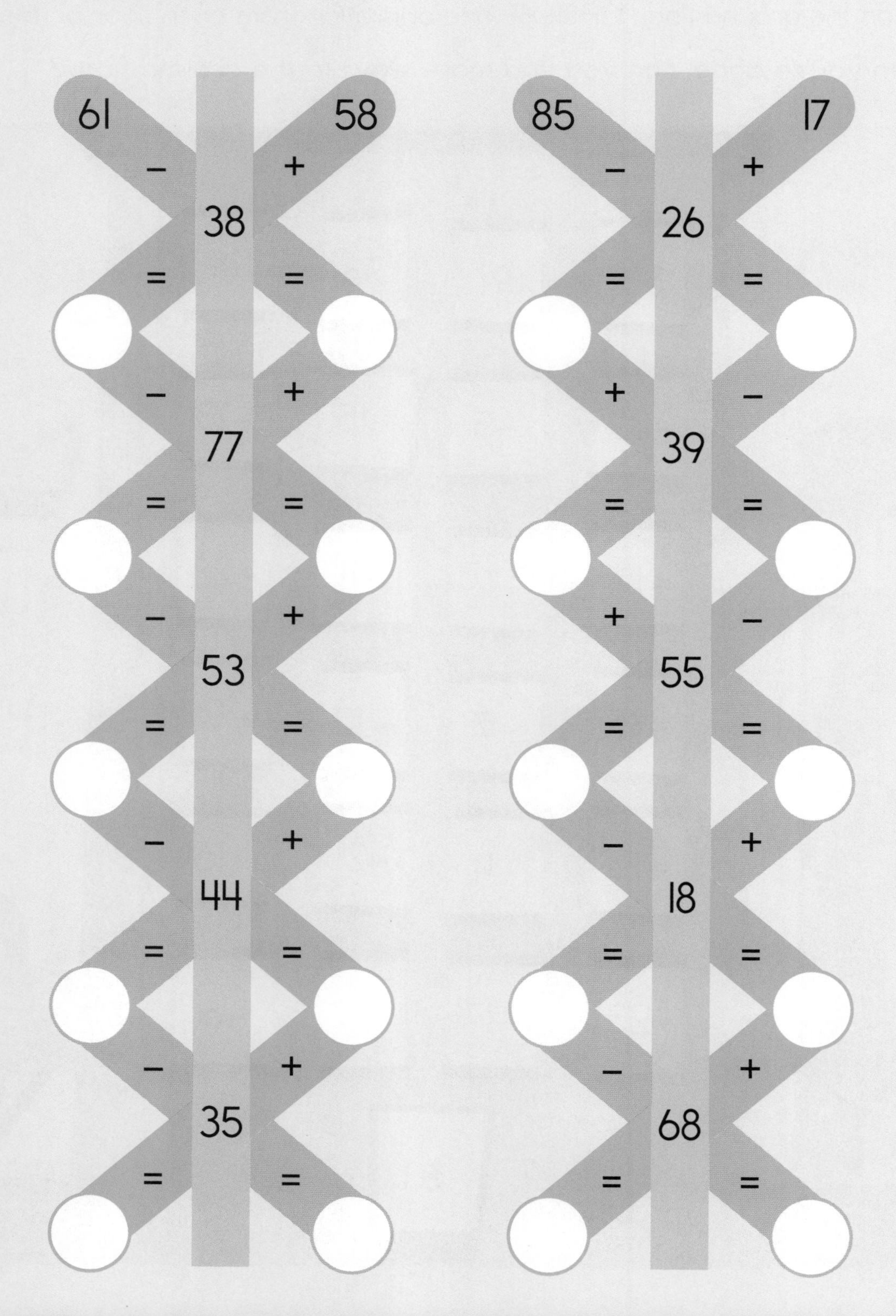

Ground Floor

DRAW a line connecting one of the numbers in the windows at the top with the answer on the ground floor. CHOOSE one operation from each floor on the way down.

HINT: When you're done, can you find more ways to the ground floor?

Sandy Shore

DRAW two straight lines in the sand to create four equal sets of shells.

Car Clubs

FIND each kind of car in the picture. WRITE the number of groups that can be made from each kind of car.

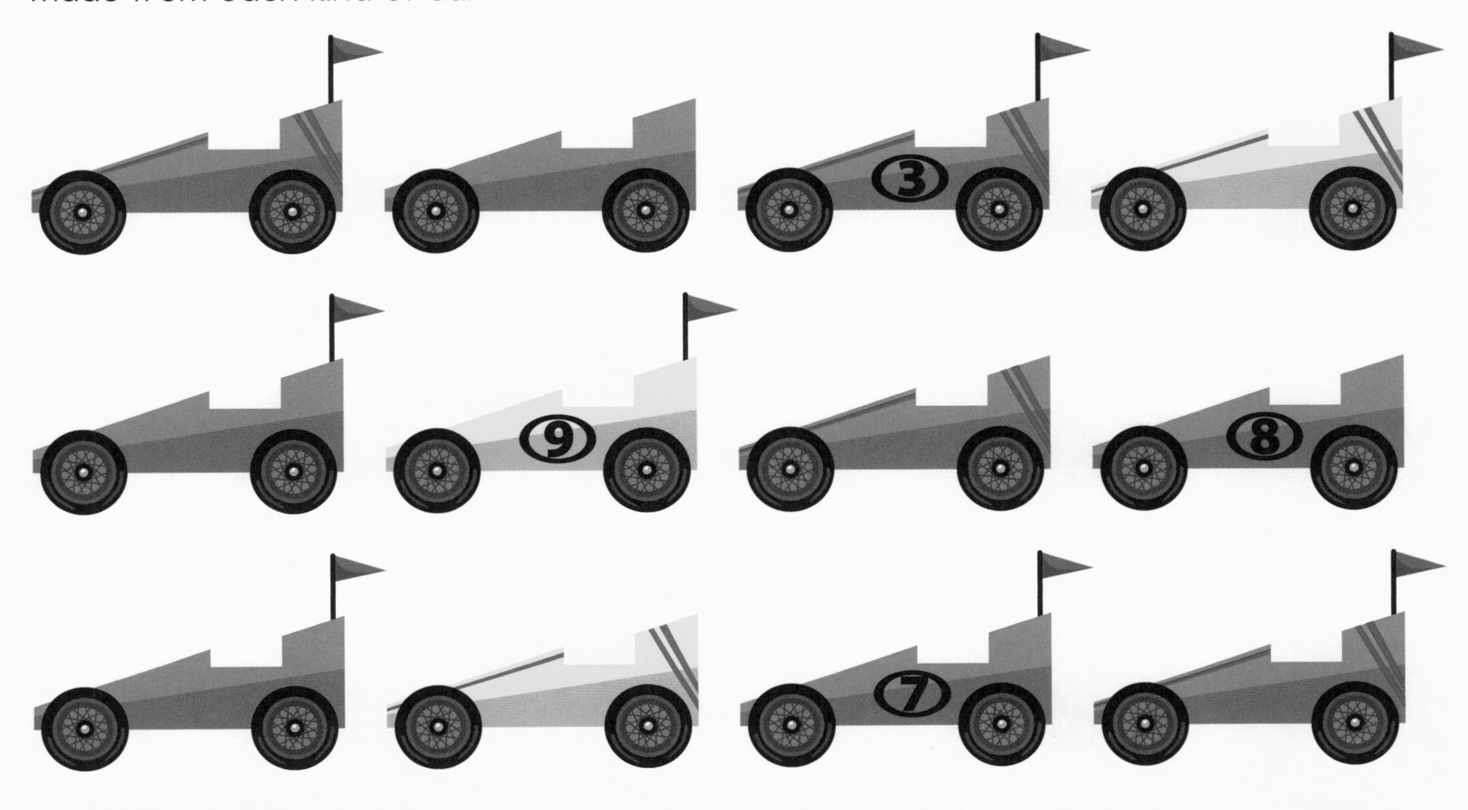

1. Groups of three blue cars 3
2. Groups of two cars with numbers ______
3. Groups of two cars with black wheels ______
4. Groups of four cars with flags ______
5. Groups of three yellow cars ______
6. Groups of two cars with red stripes ______

Sandy Shore

DRAW three straight lines in the sand to create six equal sets of shells.

Bug Buddies

FIND each kind of bug in the picture. WRITE the number of groups that can be made from each kind of bug.

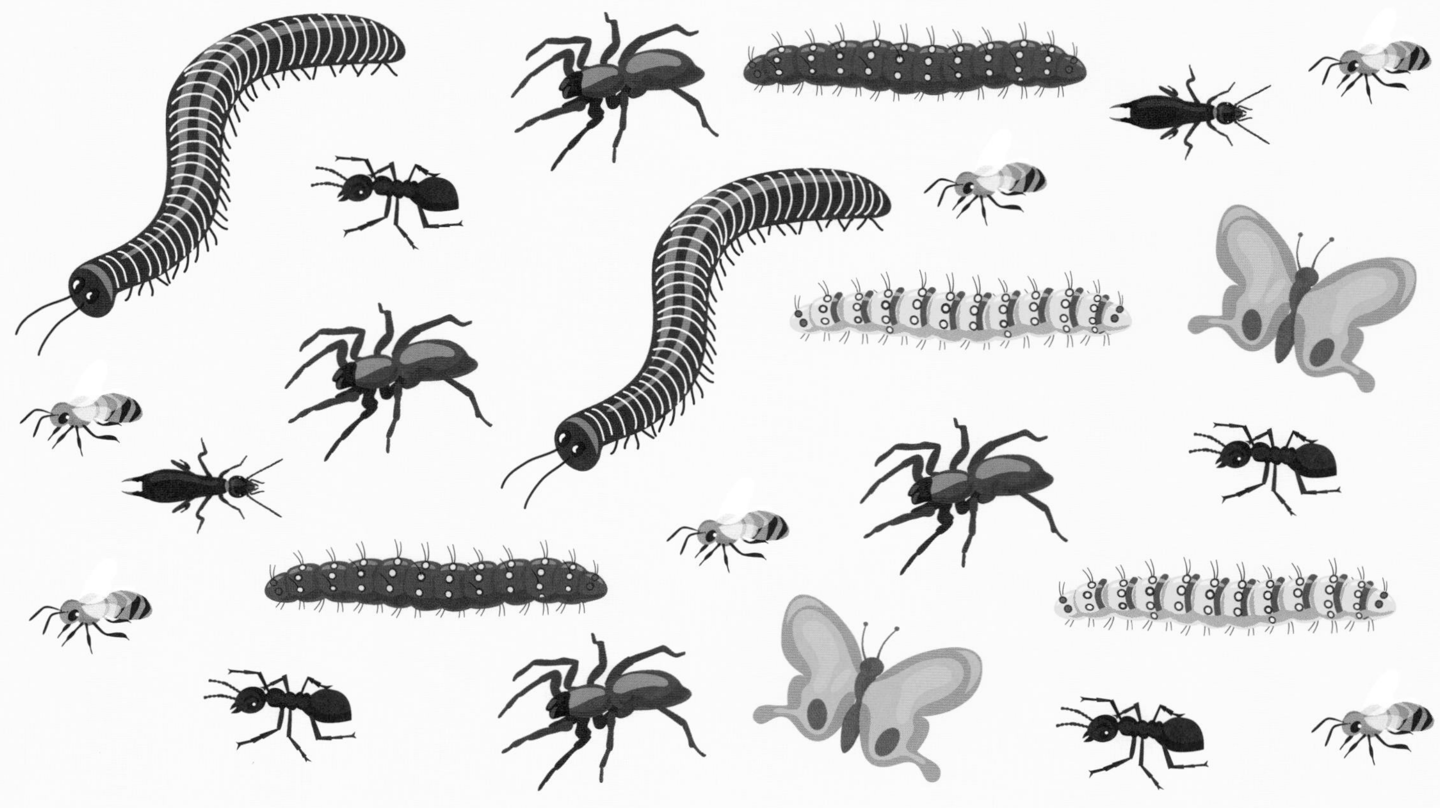

1. Groups of four brown bugs ______
2. Groups of two bugs with wings ______
3. Groups of three bugs with more than eight legs ______
4. Groups of two striped bugs ______
5. Groups of four bugs that can't fly ______
6. Groups of two bugs with eight legs _____

Bean Counter

Use the beans from page III, and SHARE them equally among the bowls. Start with the number of beans listed, then WRITE how many beans will be in each bowl when the beans are shared equally. (Save the beans to use again later in the workbook.)

1. 9 beans: ________ per bowl
2. 15 beans: ________ per bowl
3. 6 beans: ________ per bowl
4. 18 beans: ________ per bowl
5. 24 beans: ________ per bowl
6. 12 beans: ________ per bowl

Going Bananas

The monkeys want bananas! DRAW a line from each monkey at the top to his basket at the bottom to collect the bananas. You can only cross each banana once, and the monkeys must each end up with the same number of bananas. You must use all of the bananas.

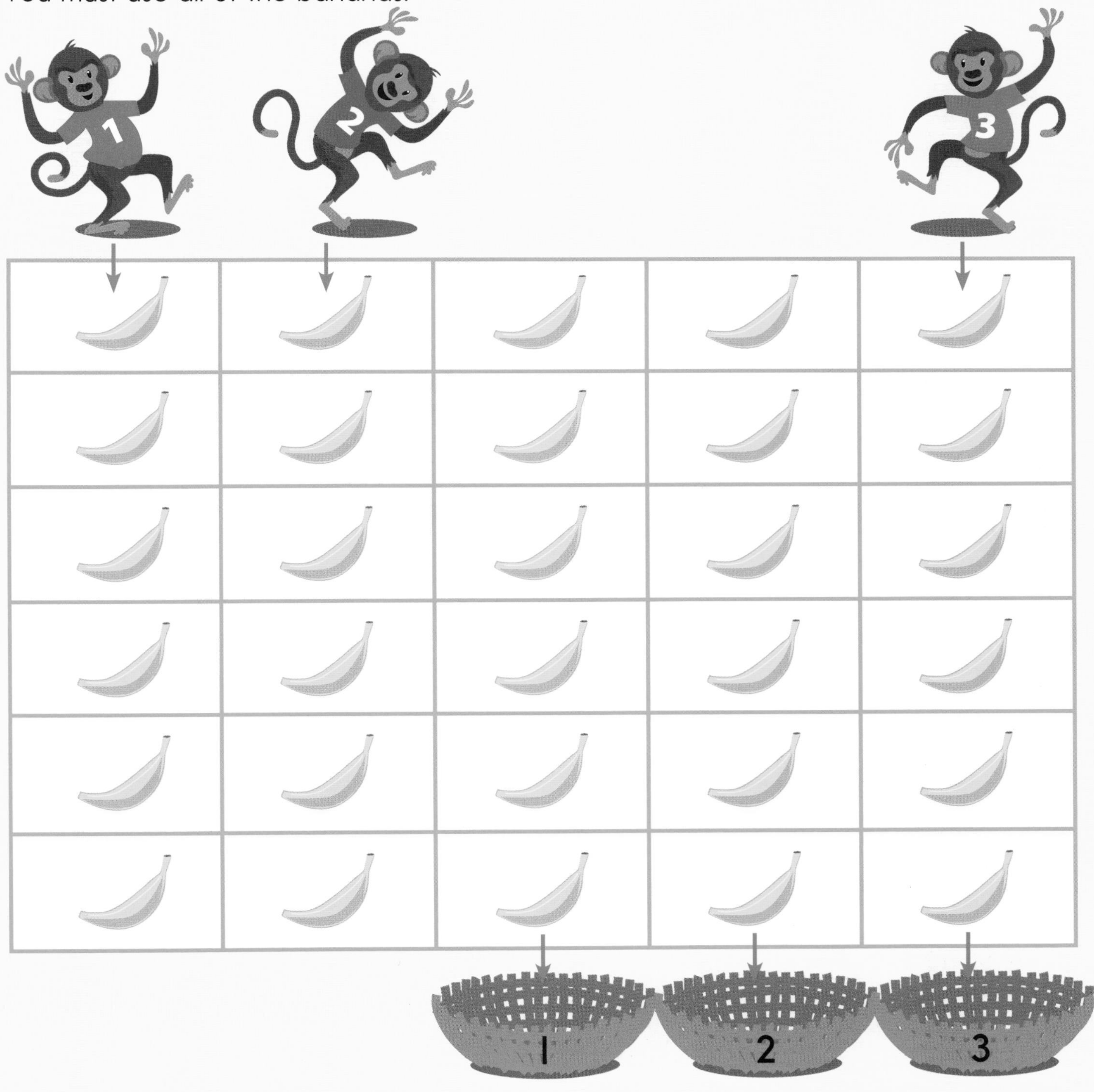

Bean Counter

Use the beans from page 111, and SHARE them equally among the bowls. Start with the number of beans listed, then WRITE how many beans will be in each bowl when the beans are shared equally. (Save the beans to use again.)

1. 16 beans: ______ per bowl
2. 32 beans: ______ per bowl
3. 8 beans: ______ per bowl
4. 28 beans: ______ per bowl
5. 20 beans: ______ per bowl
6. 48 beans: ______ per bowl

Count and Capture

The object of the game is to capture the most beans. Use the beans from page 111, an egg carton with the top cut off, and two small bowls. READ the rules. PLAY the game.

1. Set up the game by putting the bowls at either end of the egg carton and placing four beans in each egg carton section. Each player owns the sections on his side.
2. The youngest player goes first. To take a turn, a player scoops up all of the beans from one of his sections, then places one bean at a time in each section, moving to the right (counterclockwise).

   Example: Player 1 scoops up four beans from the yellow section, and places one bean in each orange section that follows to the right.

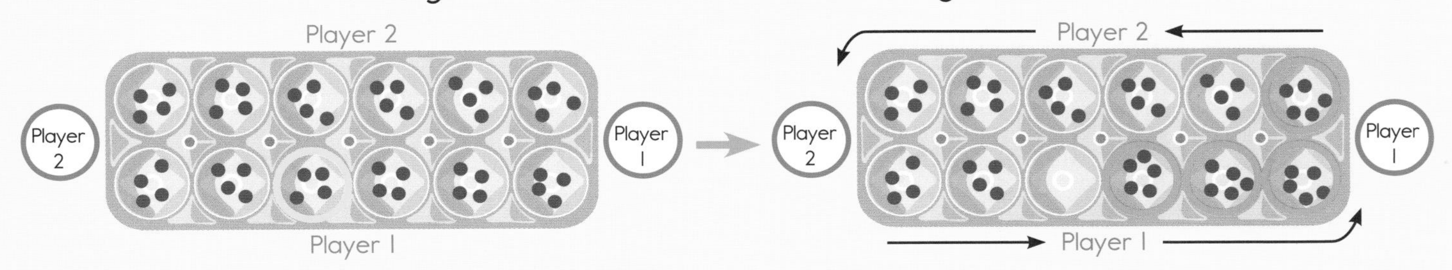

3. A player captures beans when placing a bean in any of the other player's sections to make a total of two or three beans. Captured beans go in the player's bowl.

   Example: Player 1 moves the beans from the yellow section and captures the beans in the two orange sections because the total number of beans is two and three. He then puts all of the captured beans in his bowl.

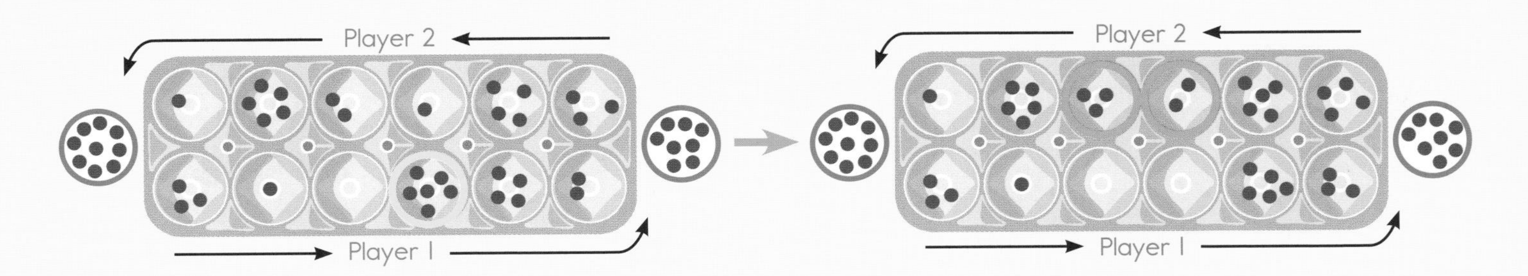

4. Play continues until all of the beans are captured or no more moves can be made.

The player with the most beans wins!

Code Breaker

FIND the odd or even number in each row. Then WRITE the letter that matches each number to solve the riddle.

FIND the odd number in each row.

1.	3	4	8	K
2.	10	7	6	W
3.	2	9	14	A
4.	8	12	1	H

FIND the even number in each row.

5.	7	5	6	E
6.	12	9	11	T
7.	8	15	3	Y
8.	1	10	13	S

How do you make seven even?

____ ____ ____ ____   ____ ____ ____ ____
12 9 3 6   9 7 9 8

____ ____ ____   ____.
12 1 6   10

Odd Way Out

START at the arrow. DRAW a line through only odd numbers to get to the smiley face.

Where's My Brain?

START at the arrow. DRAW a path through only even numbers to reach the brain.

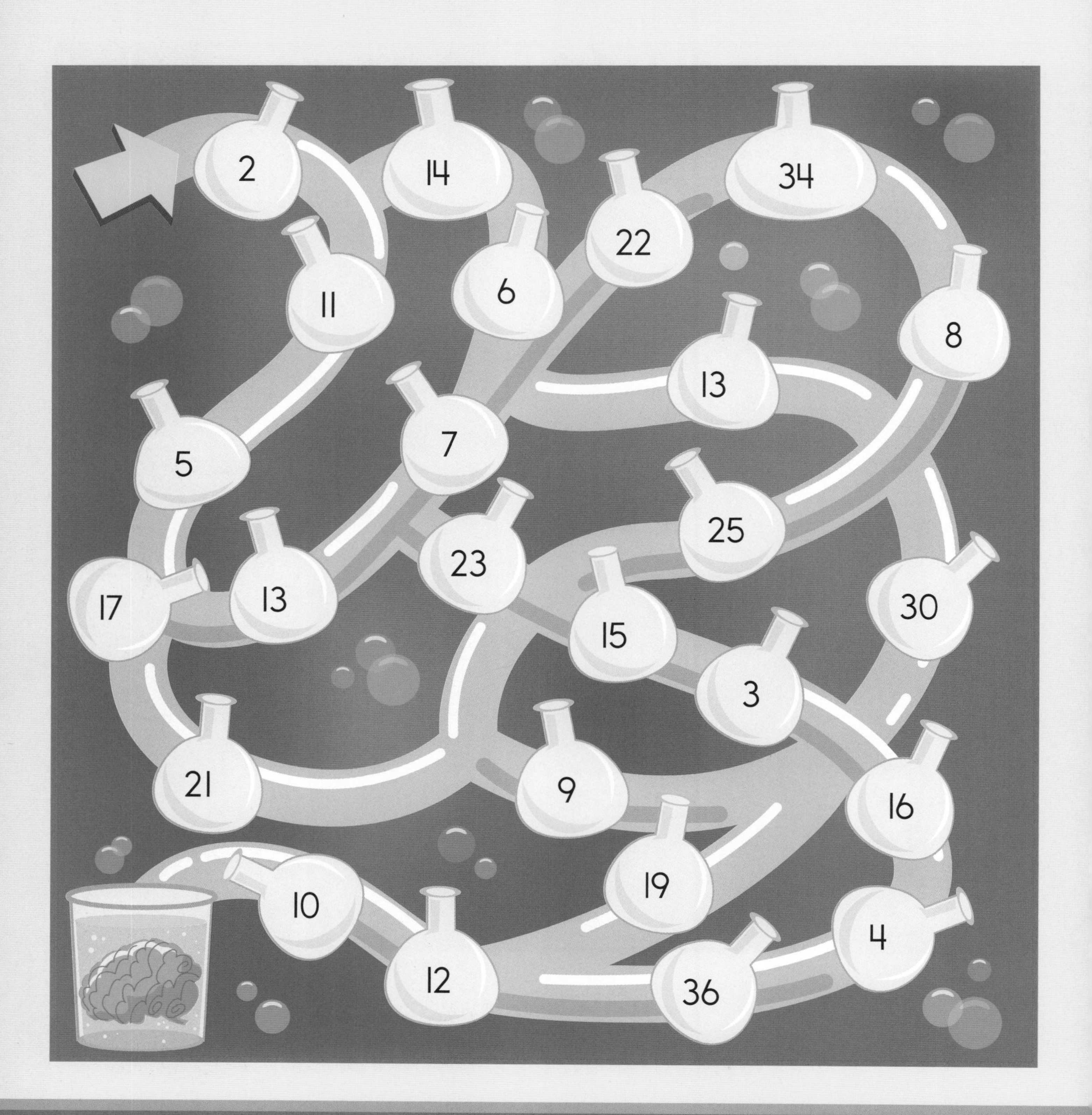

Alien Adventure

Can you be the first alien to reach the spaceship? Use two small objects as playing pieces and the spinner from page 113. READ the rules. PLAY the game! (Save the spinner to use again later in the workbook.)

Rules: Two players

1. Place the playing pieces at Start.
2. Take turns spinning the spinner. If you spin Odd, move to the next odd number. If you spin Even, move to the next even number.
3. If you land on a space with an asteroid, you lose a turn.

The first player to get to the spaceship wins!

Button Up

DRAW four straight lines to create nine equal sets of buttons.

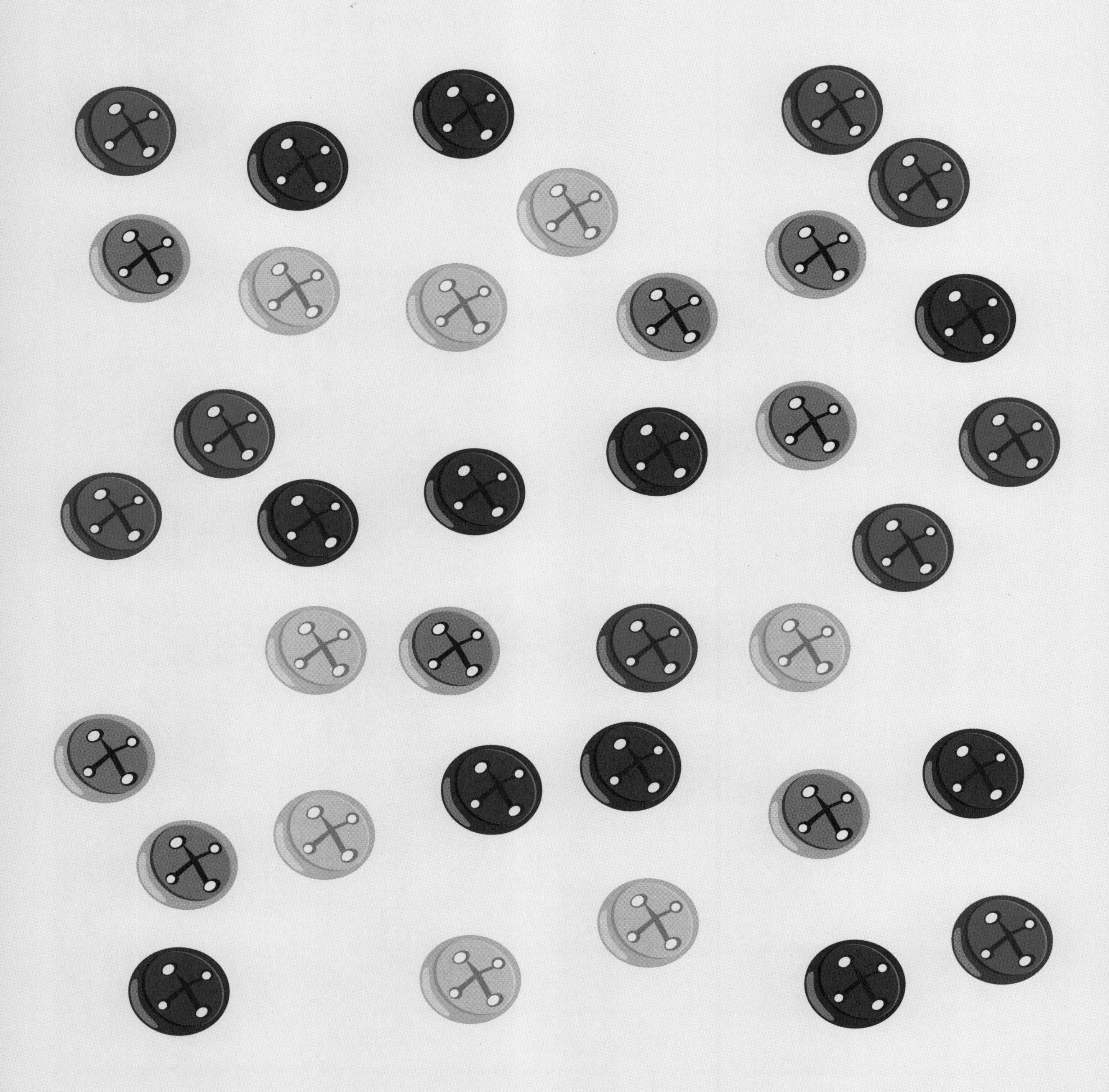

Odd Way Out

START at the arrow. DRAW a line through only odd numbers to get to the end.

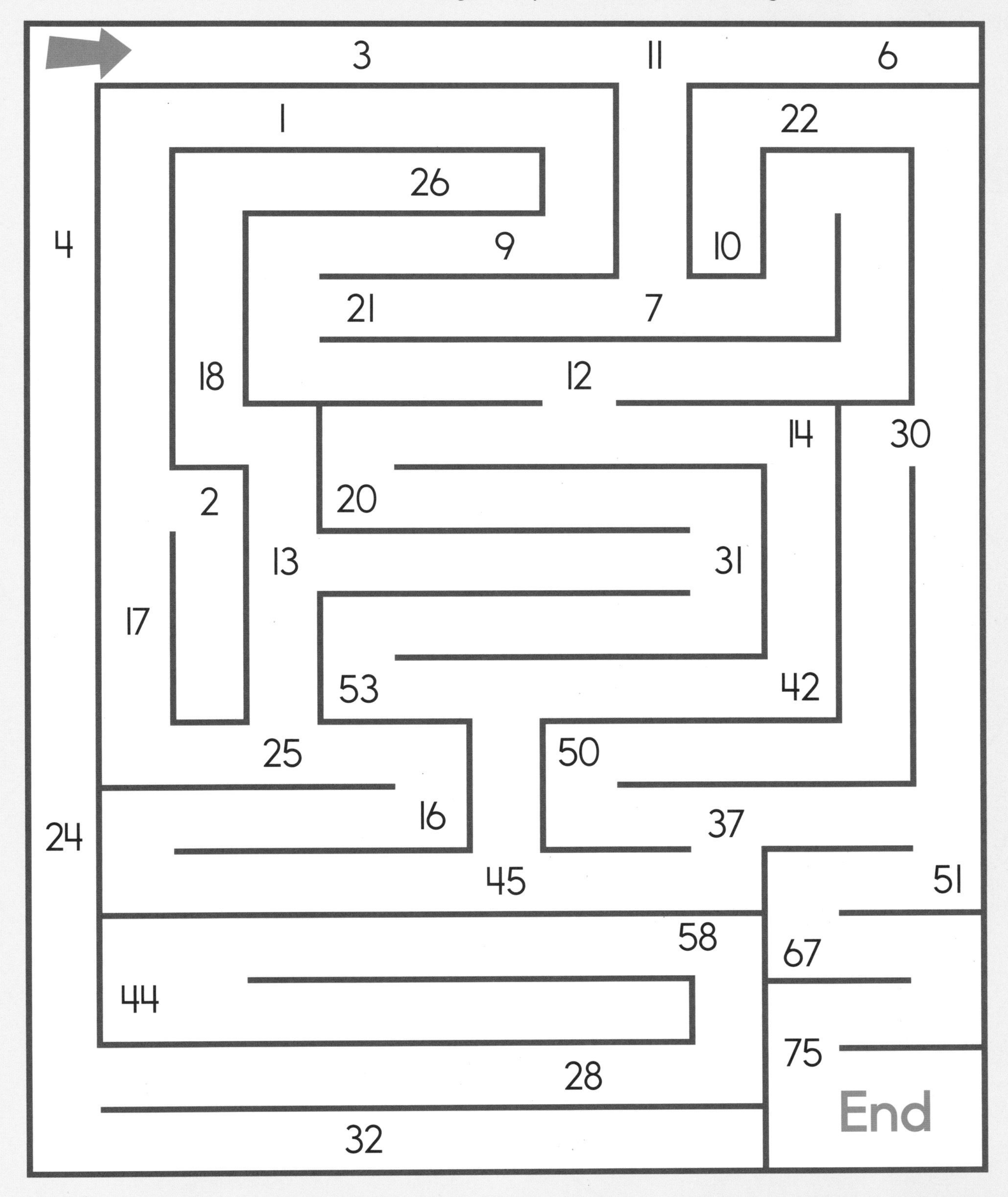

Totally Tangled

FIND the fraction and picture pairs that are connected, and COLOR any fraction that does **not** match the picture.

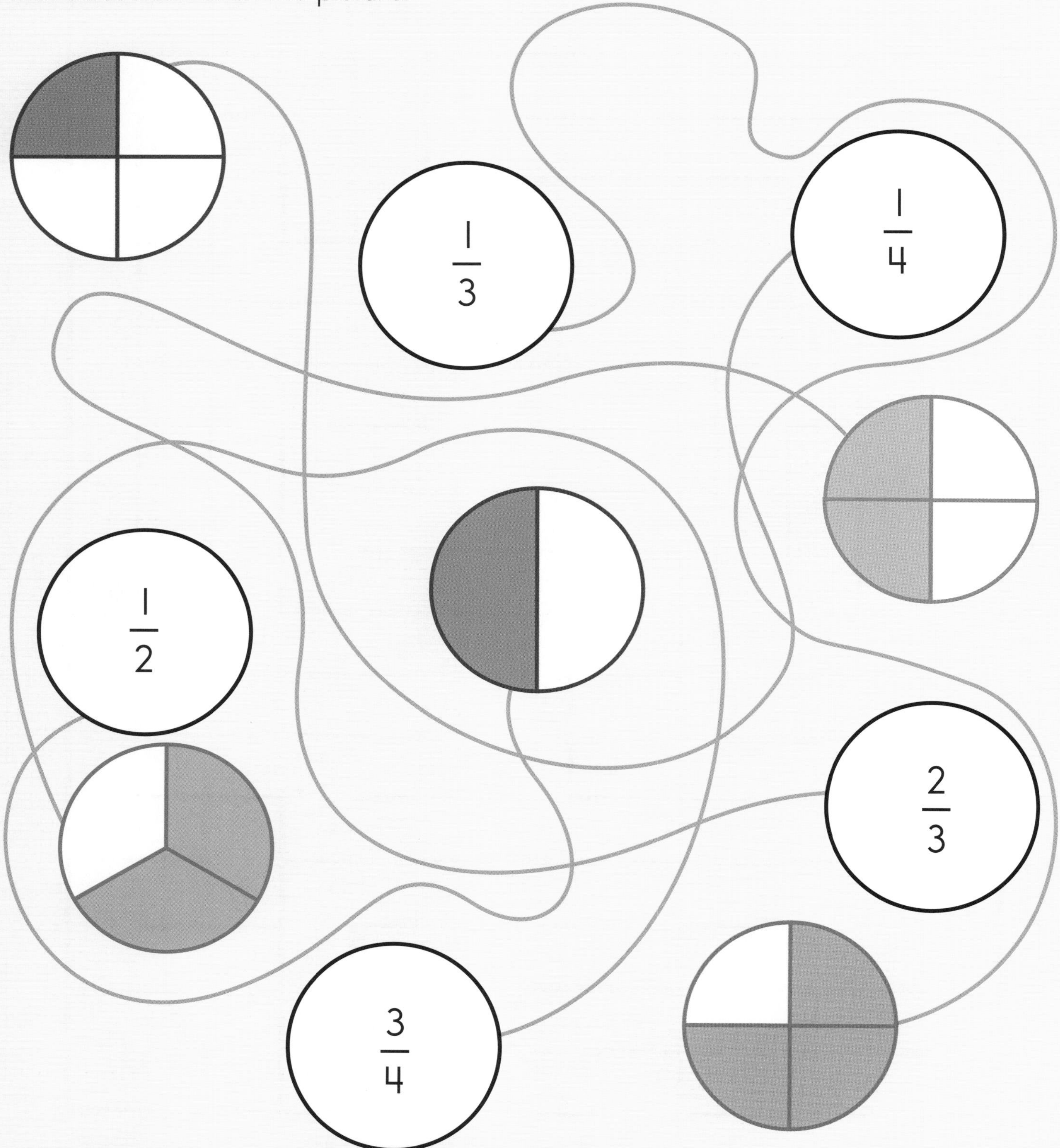

Make a Match

CUT OUT the fractions and pictures. READ the rules. PLAY the game!

Rules: Two players

1. Place the cards face down on a table.
2. Take turns turning over two cards at a time.
3. Keep the cards when you match a fraction and a picture that shows that fraction shaded.

The player with the most matches wins!

$\frac{1}{4}$		$\frac{1}{3}$	
$\frac{1}{2}$		$\frac{2}{3}$	
$\frac{2}{2}$		$\frac{3}{4}$	

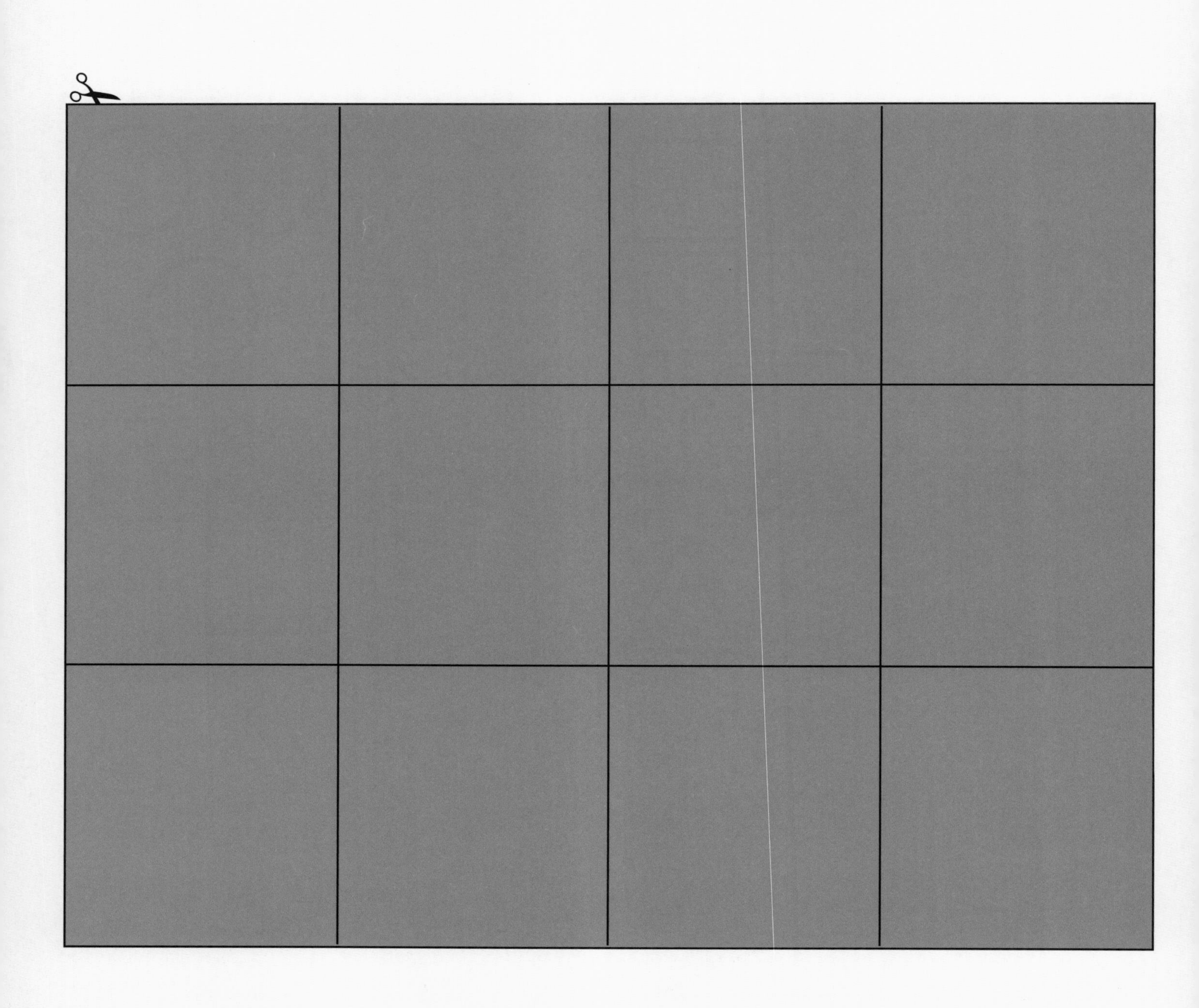

Mystery Picture

COLOR each section according to the fractions to reveal the mystery picture.

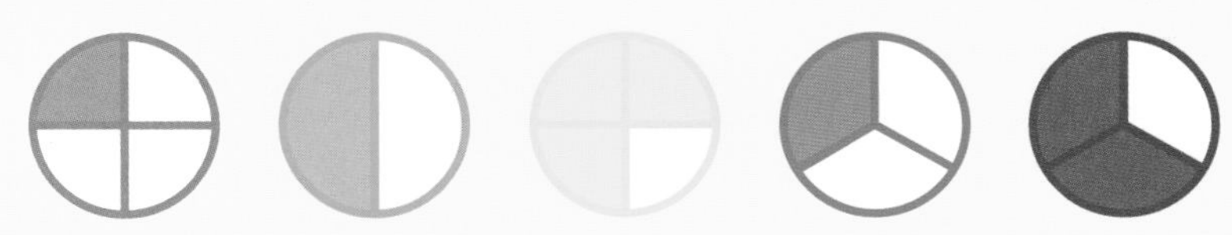

Fraction Factory

Can you be the first to reach the end? Use two small objects as playing pieces and the spinner from page 114. READ the rules. PLAY the game!

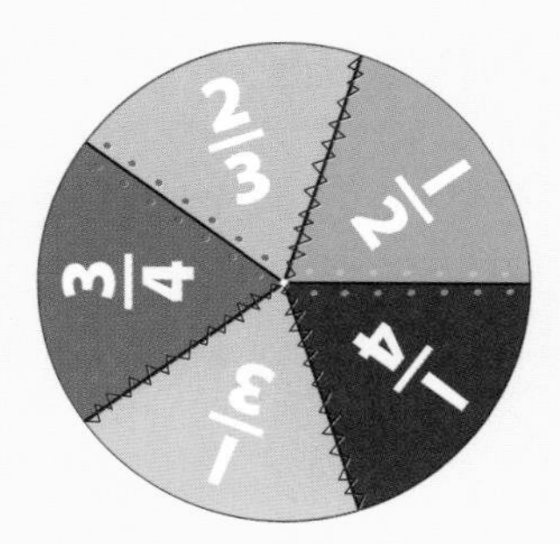

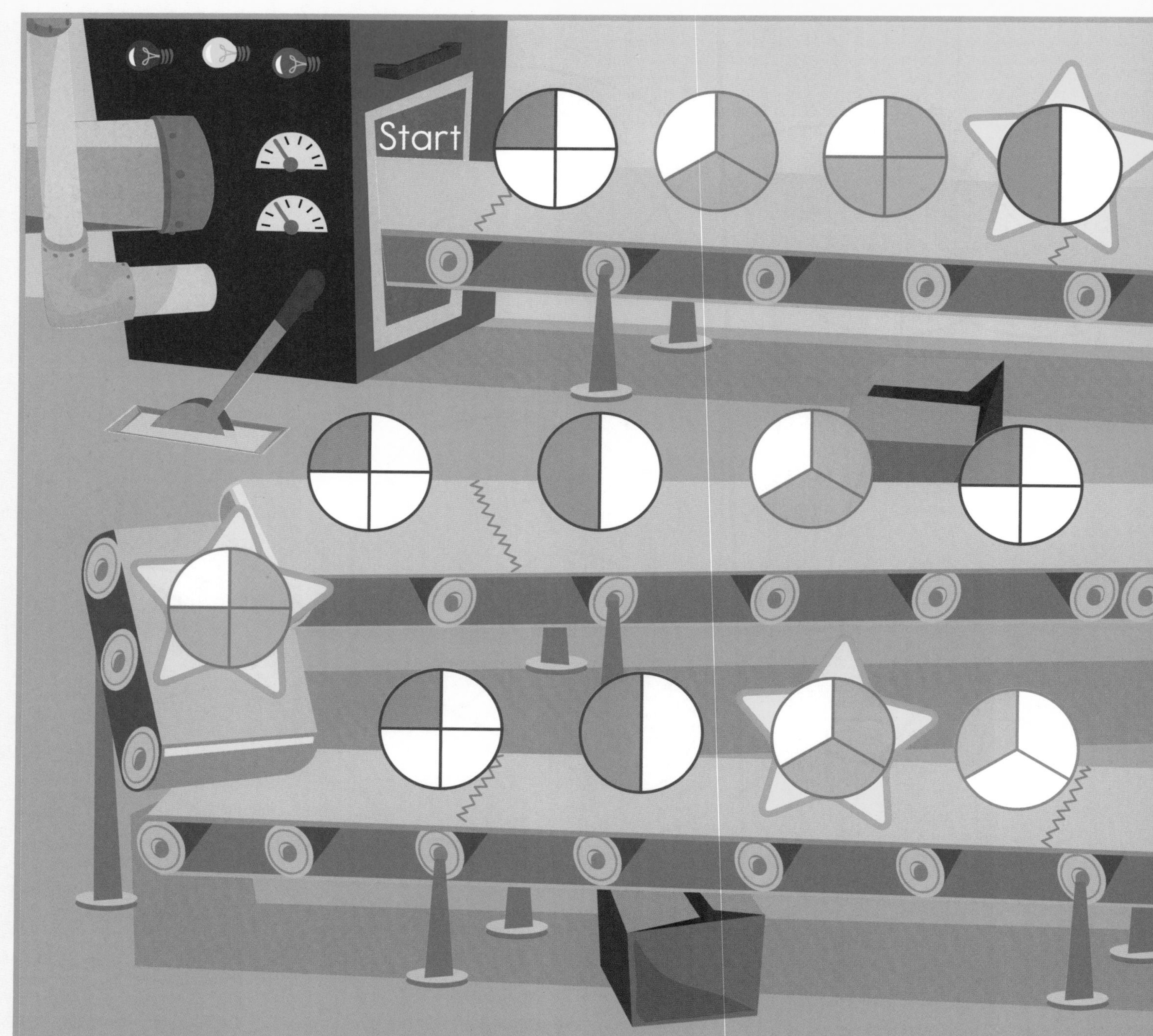

Fractions

Rules: Two players

1. Place the playing pieces at Start.
2. Take turns spinning the spinner. Move to the closest fraction picture that matches the fraction on the spinner.
3. If you land on a space with a star, you get to spin again.

The first player to get to the End box wins!

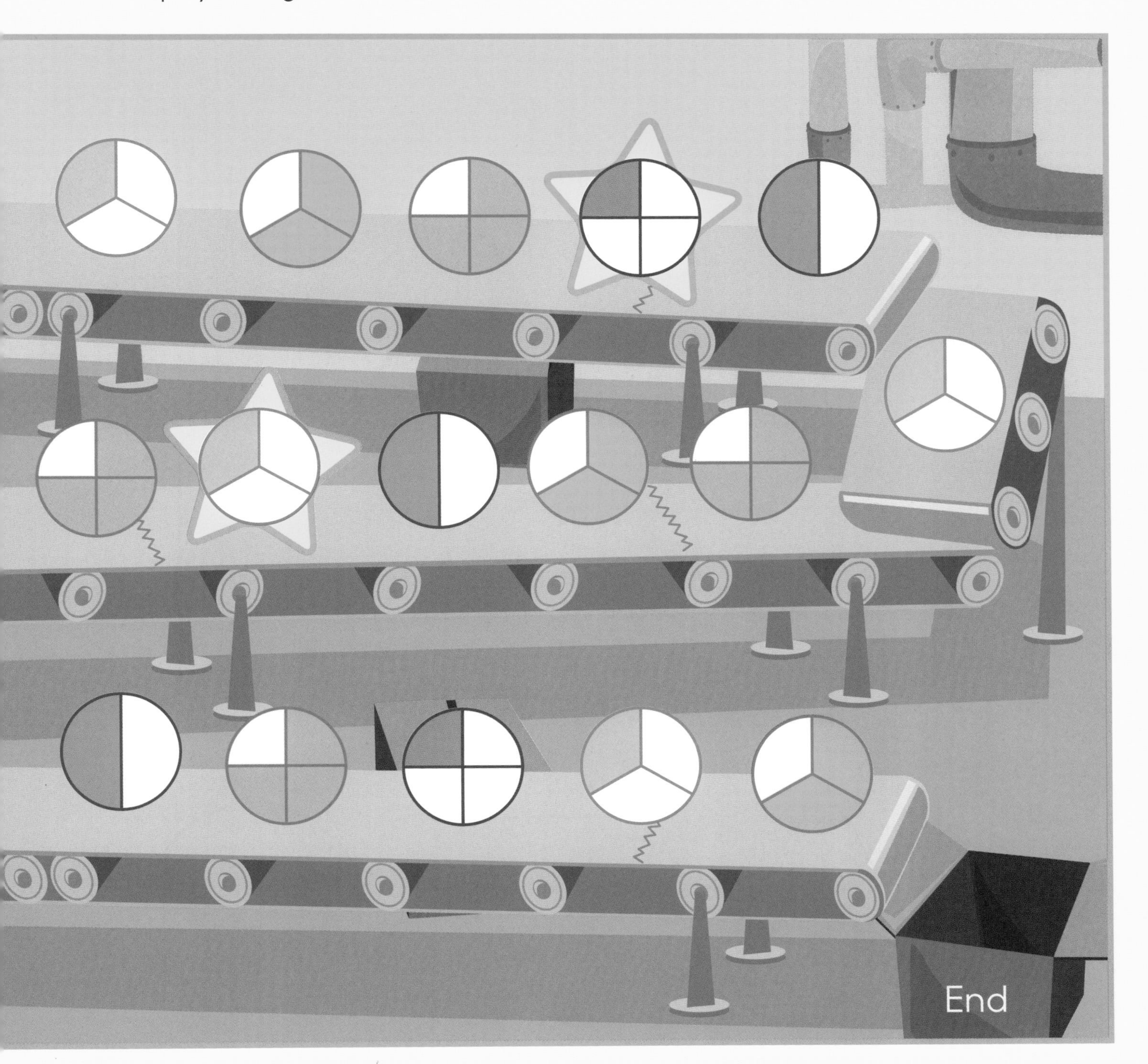

Picking Pairs

DRAW a line to connect each equivalent pair of fraction pictures.

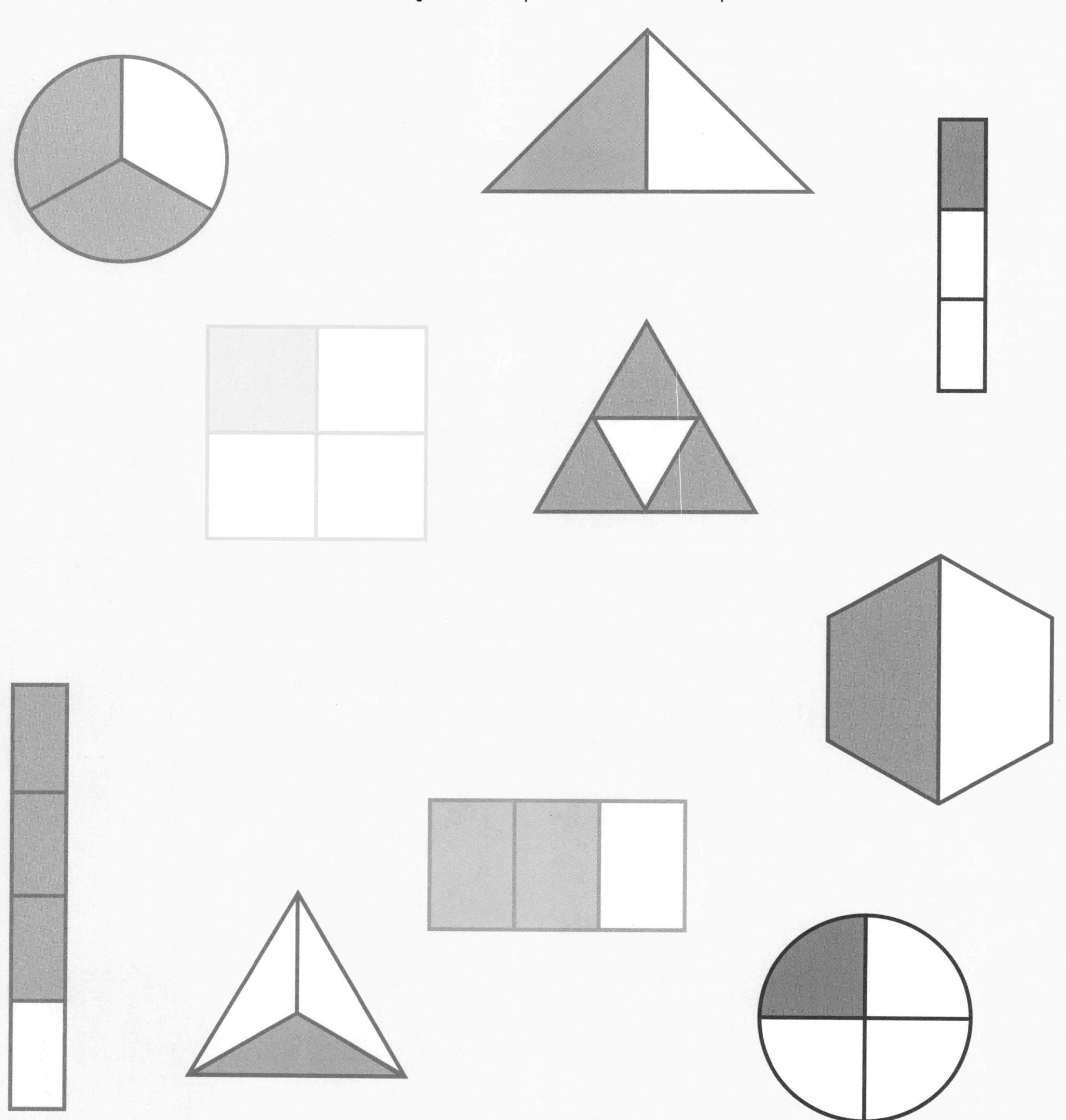

Totally Tangled

Each fraction is connected to another fraction. FIND the pairs of fractions, and COLOR the circle with the larger fraction.

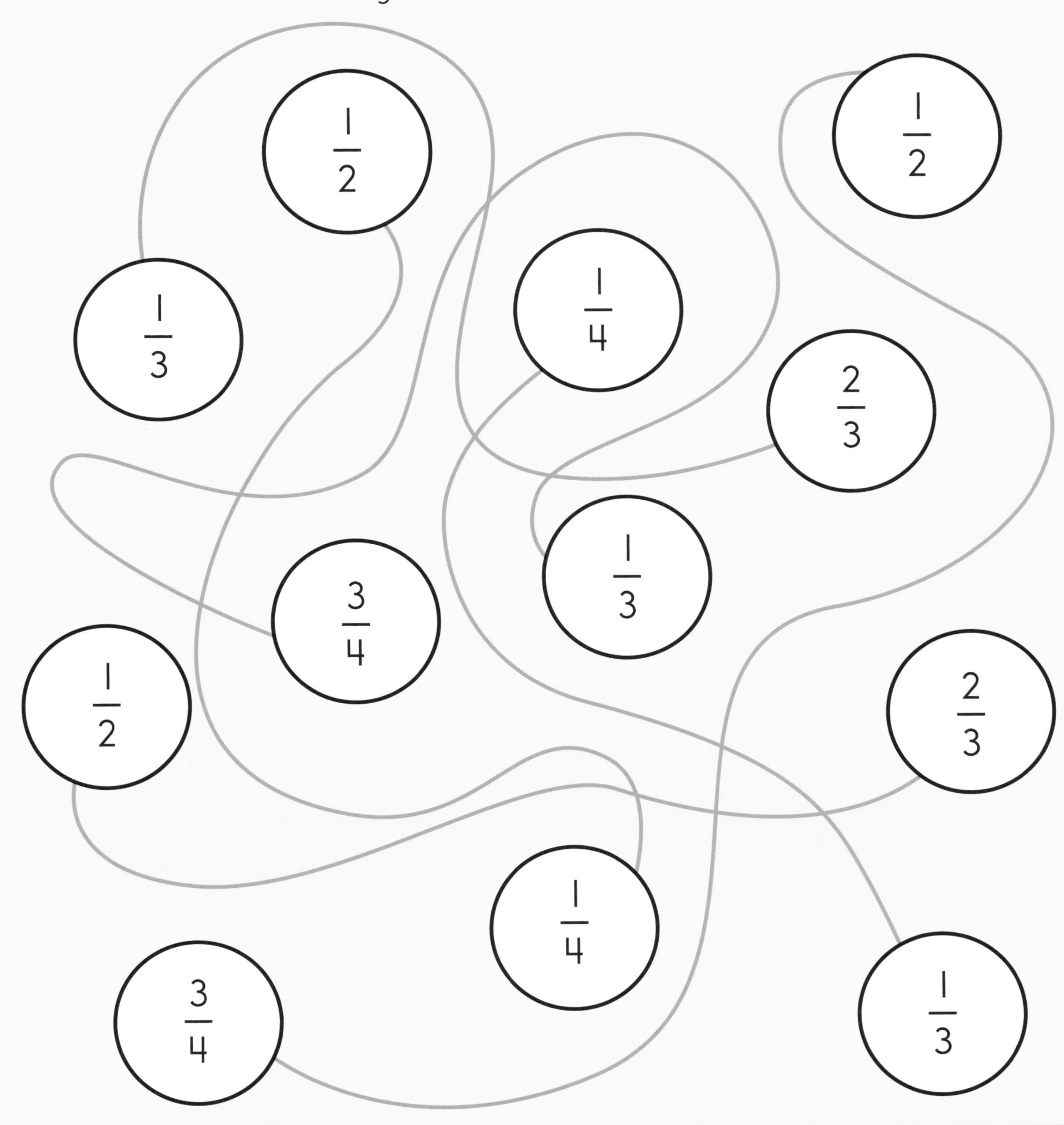

Just Right

WRITE each of these fractions next to a smaller fraction picture.

HINT: There may be more than one place to put a fraction, but you need to use every fraction.

$\frac{1}{3}$ $\frac{3}{4}$ $\frac{2}{3}$ $\frac{1}{2}$ $\frac{4}{4}$

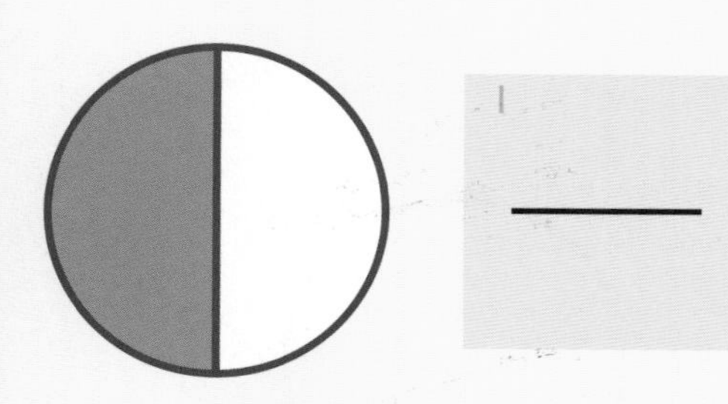

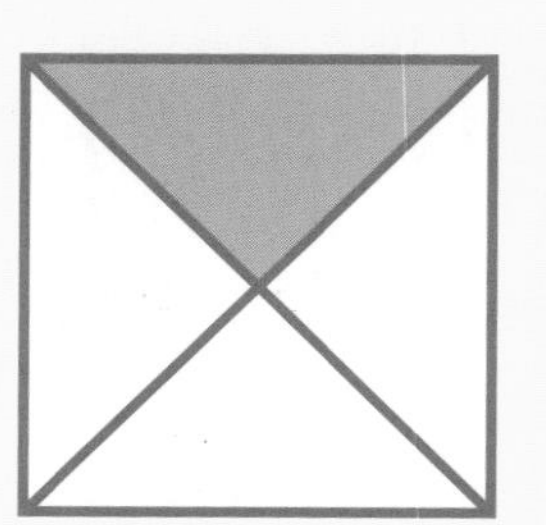

2 _____

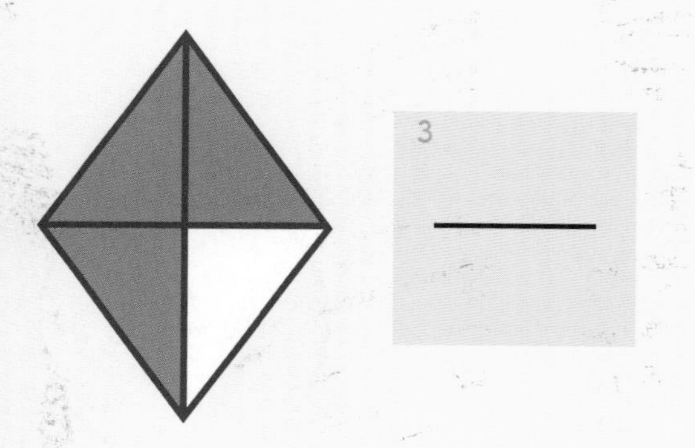

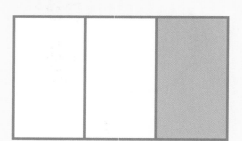

4 _____

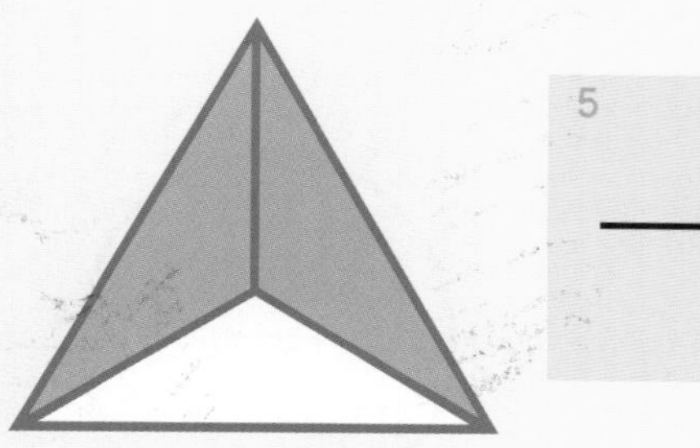

Code Breaker

CIRCLE the larger fraction in each pair. Then WRITE the letter that matches each fraction to solve the riddle.

1		2		3		4	
$\frac{1}{4}$	$\frac{1}{3}$	$\frac{3}{4}$	$\frac{2}{4}$	$\frac{1}{2}$	$\frac{2}{2}$	$\frac{1}{2}$	$\frac{1}{3}$
W		E		T		O	

5		6		7		8	
$\frac{2}{4}$	$\frac{1}{3}$	$\frac{3}{3}$	$\frac{1}{2}$	$\frac{2}{4}$	$\frac{2}{3}$	$\frac{4}{4}$	$\frac{2}{3}$
L		H		I		N	

Why did the boat carrying three thirds sink?

There was a ____ ____ ____ ____ ____

$\frac{1}{3}$ $\frac{3}{3}$ $\frac{1}{2}$ $\frac{2}{4}$ $\frac{3}{4}$

____ ____ ____ ____.

$\frac{2}{3}$ $\frac{4}{4}$ $\frac{2}{3}$ $\frac{2}{2}$

What's the Password?

WRITE the letters that form a fraction of each word. Then WRITE the letters in order to find the secret password.

The first $\frac{1}{2}$ of PECK ________ 1

The last $\frac{1}{4}$ of RAIN ________ 2

The first $\frac{1}{3}$ of COW ________ 3

The middle $\frac{1}{3}$ of CHILLY ________ 4

The last $\frac{1}{4}$ of TAKE ________ 5

The second $\frac{1}{2}$ of OR ________ 6

The first $\frac{2}{3}$ of ASK ________ 7

The last $\frac{1}{3}$ of SINGER ________ 8

Password: ______ ______ ______ ______ ______ ______

______ ______ ______ ______ ______ ______

Who Am I?

READ the clues, and CIRCLE the mystery number.

HINT: Cross out any fraction picture that does not match the clues.

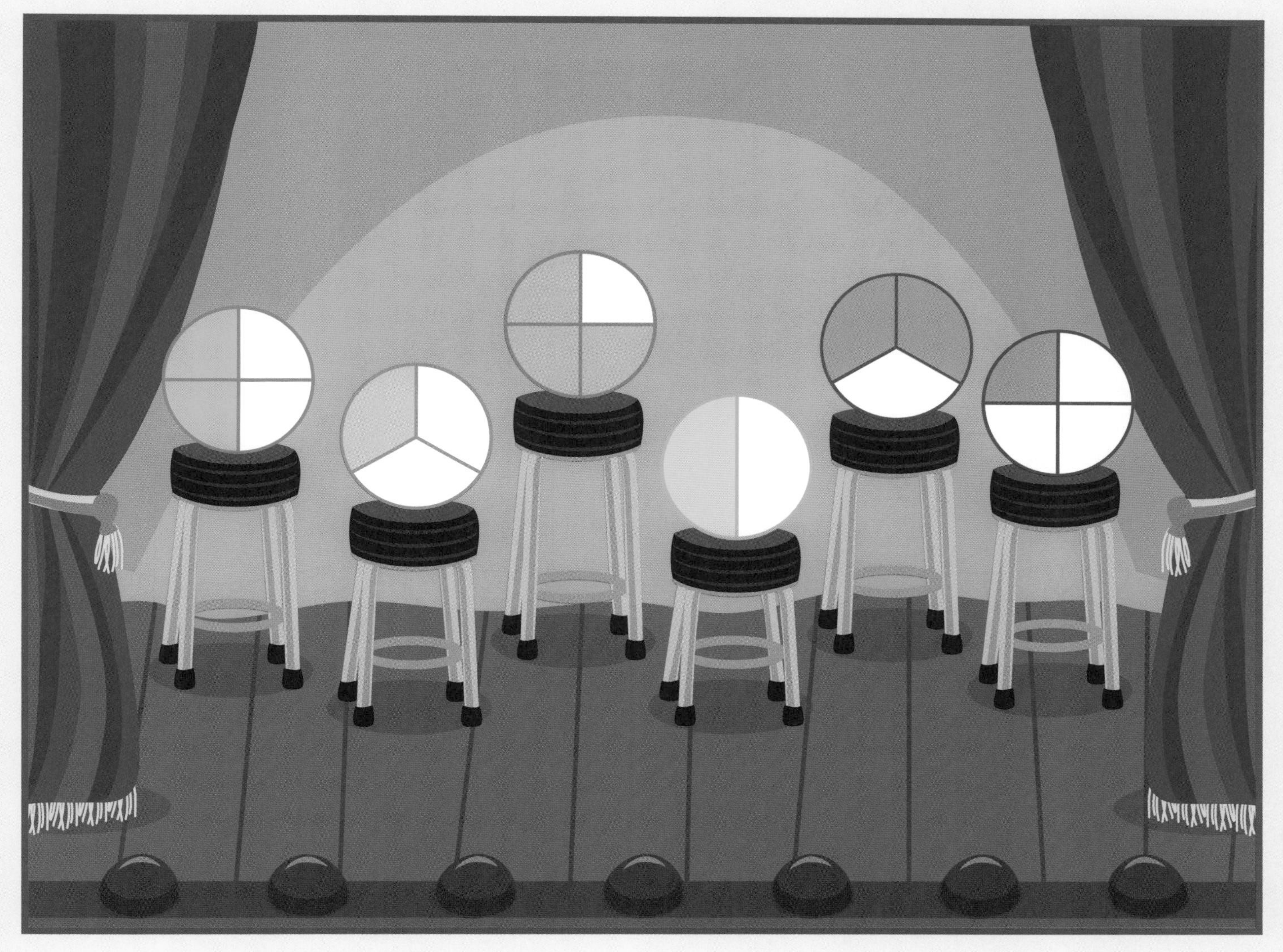

I am larger than $\frac{1}{4}$.

I am smaller than $\frac{3}{4}$.

I have two parts shaded.

I am not equivalent to anyone else.

Who am I?

Hidden Shapes

FIND each shape hidden in the picture. DRAW a line to connect each shape with its location in the picture.

HINT: Be sure to use shapes that match in size.

Doodle Pad

TRACE the shapes. Then DRAW a picture using each shape.

Shape Shifters

A shape has **symmetry** if a line can divide the shape so each half is a mirror image of the other. Use the pattern block pieces from page 115, and PLACE the pieces to make each picture symmetrical without overlapping any pieces. (Save the pattern block pieces to use again later in the workbook.)

Cool Kaleidoscope

COLOR the kaleidoscope so it is symmetrical.

HINT: Make a mirror image across each line.

Shape Shifters

Use the pattern block pieces from page 115, and PLACE the pieces to completely fill each shape without overlapping any pieces. See if you can solve the puzzles different ways. (Save the pattern block pieces to use again.)

10

Puzzling Pentominoes

Perimeter is the distance around a two-dimensional shape. Use the pentomino pieces from page 117, and PLACE the pieces to completely fill each shape without overlapping any pieces. Then WRITE the perimeter of each shape. (Save the pentomino pieces to use again.)

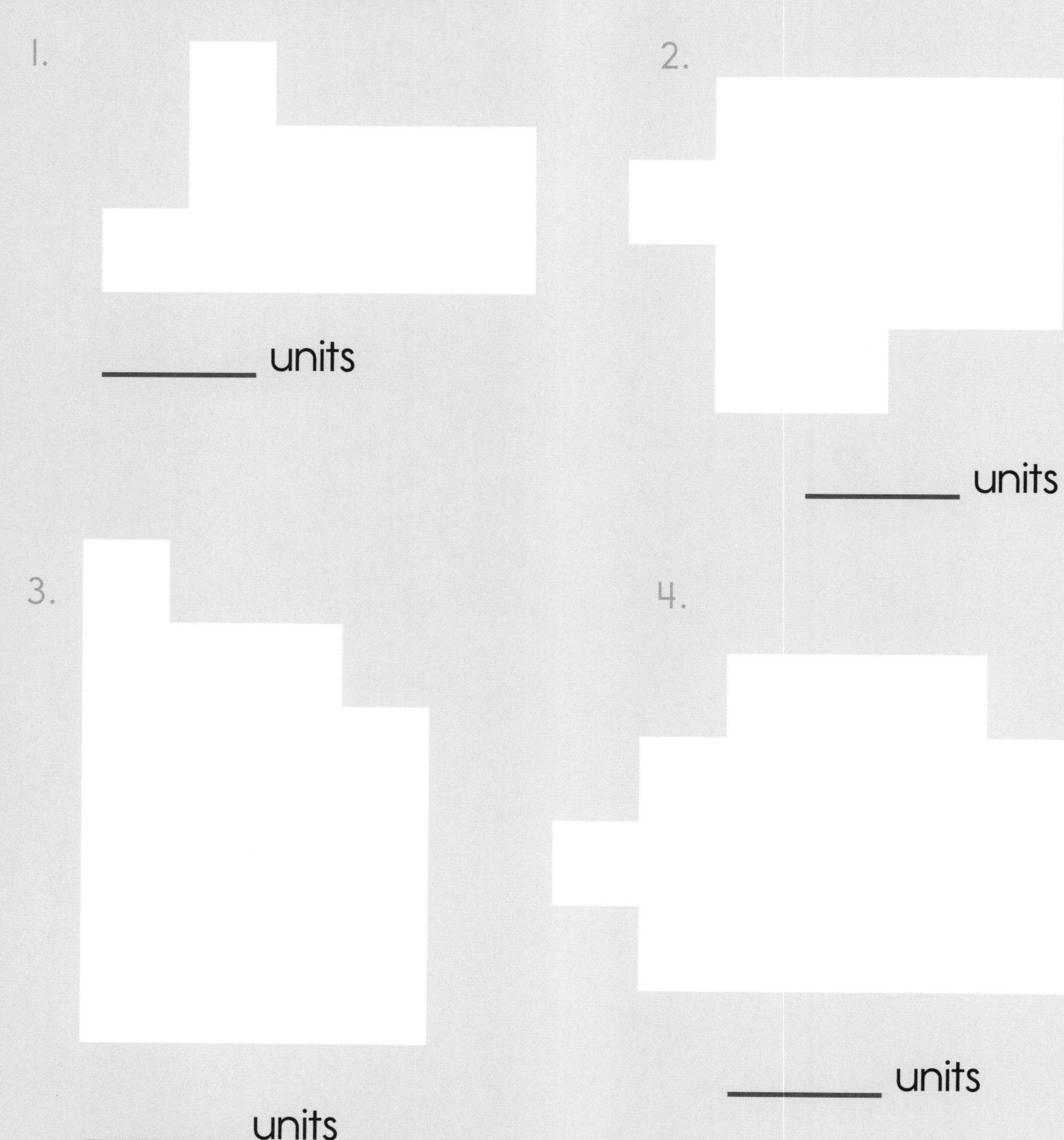

Shape Creator

DRAW six different shapes that all have a perimeter of 12 units.

Puzzling Pentominoes

Area is the size of the surface of a shape, and it is measured in square units. Use the pentomino pieces from page 117, and PLACE the pieces to completely fill each shape without overlapping any pieces. Then WRITE the area of each shape. (Save the pentomino pieces to use again.)

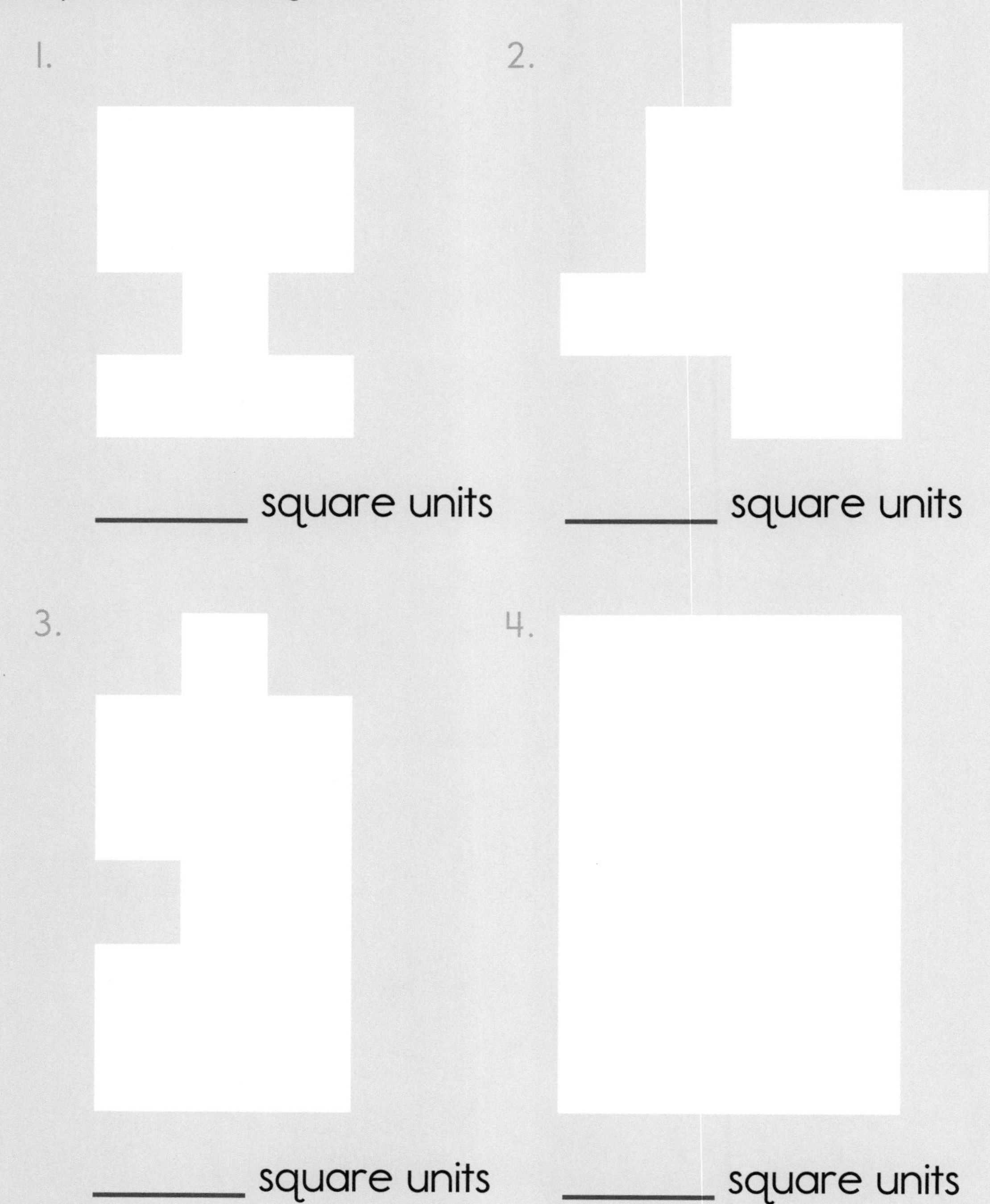

Shape Creator

DRAW six different shapes that all have an area of 10 square units.

Animal Adventure

WRITE the name of the animal that can be found at each location on the map.

HINT: Follow the letter and the number and see where the two lines meet.

1. C5 ______________________
2. C1 ______________________
3. E1 ______________________
4. D3 ______________________
5. A6 ______________________
6. A4 ______________________
7. C7 ______________________
8. B2 ______________________
9. E5 ______________________
10. E7 ______________________

Treasure Hunt

Use the squares on the map to find the pirate treasure. FOLLOW the directions. DRAW an X where the treasure is buried.

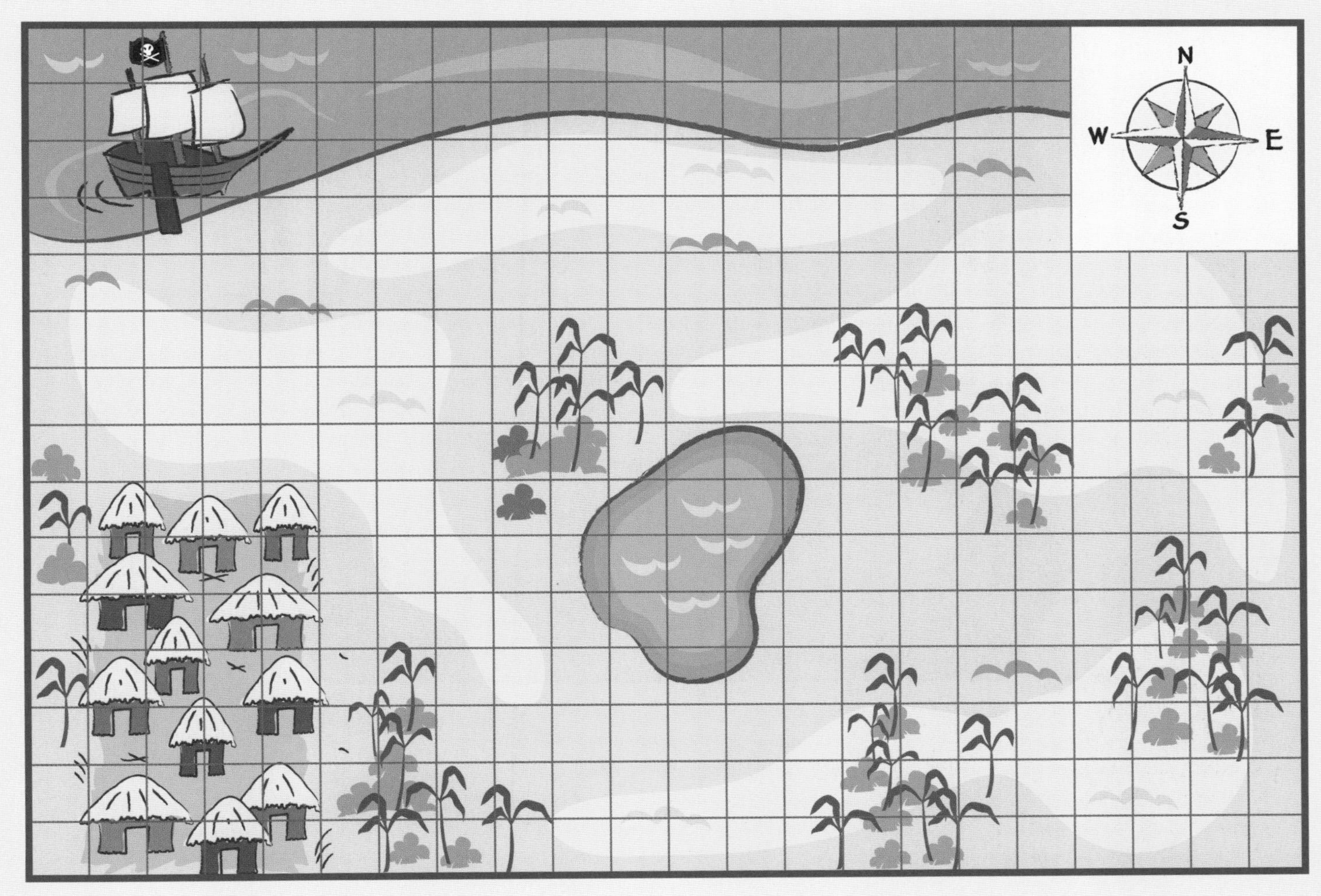

Blackbeard's Treasure

1. From the ship travel 4 squares south. Do not enter the village or the people will be curious.
2. Go 5 squares to the east, and turn to head 5 squares south.
3. Travel 6 squares to the east and 2 squares to the north to get around the lake.
4. Go another 5 squares east and you'll nearly be there.
5. Head 5 squares north. Can the treasure be near?
6. Go 2 squares west and draw an X. That's where the treasure will be!

Shape Shifters

Use the pattern block pieces from page 115, and PLACE the pieces to make the picture symmetrical without overlapping any pieces.

Puzzling Pentominoes

Use the pentomino pieces from page 117, and PLACE the pieces to completely fill each shape without overlapping any pieces.

These shapes all have the same area. WRITE the perimeter, and CIRCLE the shape with the largest perimeter.

1. Perimeter ________ units

2. Perimeter ________ units

3. Perimeter ________ units

These shapes all have the same perimeter. WRITE the area, and CIRCLE the shape with the largest area.

4. Area ________ square units

5. Area ________ square units

6. Area ________ square units

Hamster Hotel

Each hamster is four coins long. MEASURE each hamster with a line of four quarters, dimes, nickels, and pennies. When you find a match, WRITE the coin name.

Sidewalk Slugs

LINE UP dimes and MEASURE each slug. DRAW lines connecting pairs of slugs that are about the same length.

Code Ruler

WRITE the letter that matches each measurement to answer the riddle.

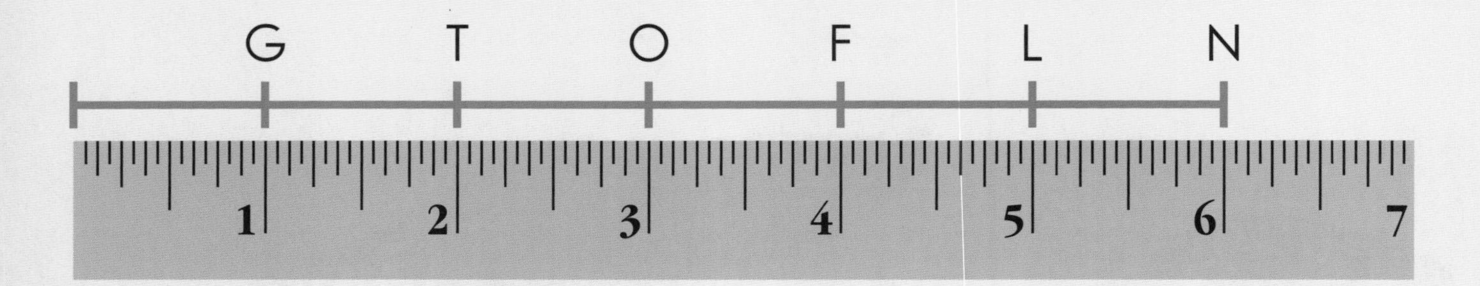

What is a ruler's favorite kind of hot dog?

A _____ _____ _____ _____ _____ _____ _____ _____ .

4 in. 3 in. 3 in. 2 in. 5 in. 3 in. 6 in. 1 in.

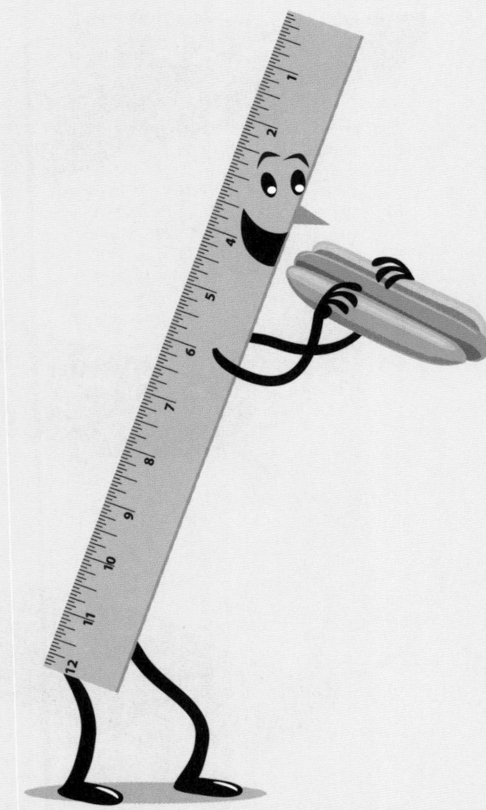

Bowl of Candy

One pack of candy in the bowl is not the same length as the others. MEASURE each pack of candy in inches, and CIRCLE the one that is not the same length.

Code Ruler

WRITE the letter that matches each measurement to answer the riddle.

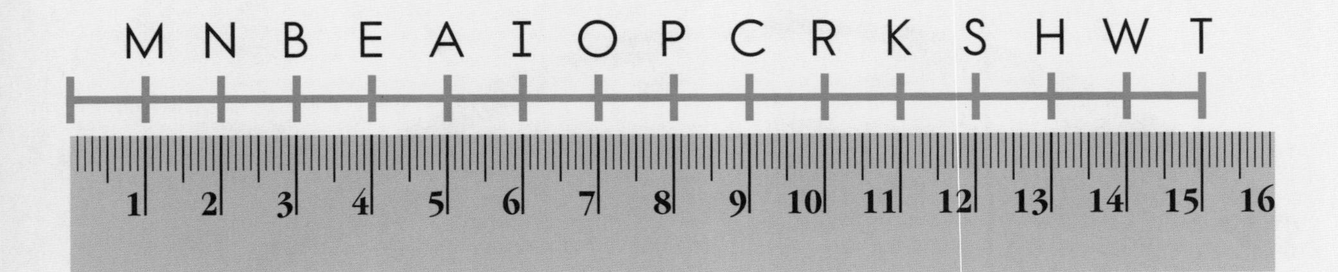

How do you measure a skunk?

____ ____   ____ ____ ____ ____ ____ ____ -

6 cm 2 cm   12 cm 9 cm 4 cm 2 cm 15 m 6 cm

____ ____ ____ ____ ____ ____ .

1 cm 4 cm 15 cm 4 cm 10 cm 12 cm

Pick a Pencil

One colored pencil is not the same length as the others. MEASURE each pencil in centimeters, and CIRCLE the one that is not the same length.

Minigolf

WRITE the numbers 1 through 6 on the golf balls so that 1 is the ball you think is closest to the hole and 6 is the golf ball you think is farthest away. Then MEASURE in inches to see if you're correct.

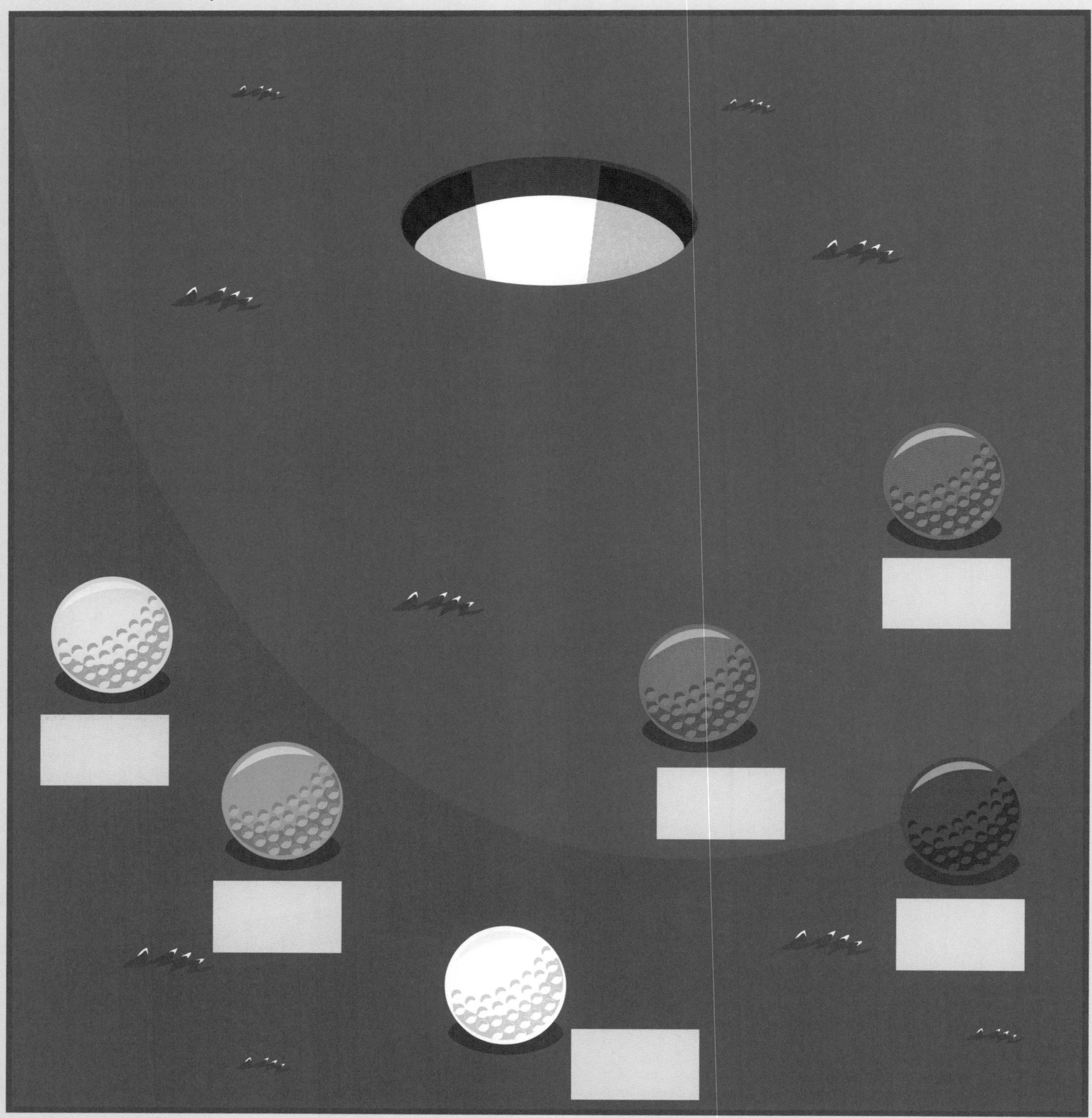

Don't Go Over

GUESS the height of each ice pop and stick in centimeters. Then MEASURE each ice pop. For every centimeter in the difference between the two measurements, COLOR a section in the white ice pop. If you get through the whole page without filling the ice pop, you win!

HINT: To find the difference, subtract the smaller measurement from the larger measurement.

So Far Away

WRITE the numbers 1 through 8 next to the ants so that 1 is the ant you think is closest to the entrance of the anthill and 8 is the ant you think is farthest away. Then MEASURE in centimeters to see if you're correct.

HINT: Use the dots to help you measure.

Don't Go Over

GUESS the distance between each matching pair of marbles in centimeters. Then MEASURE the distance. For every centimeter in the difference between the two measurements, COLOR a section in the white marble. If you get through the whole page without filling the marble, you win!

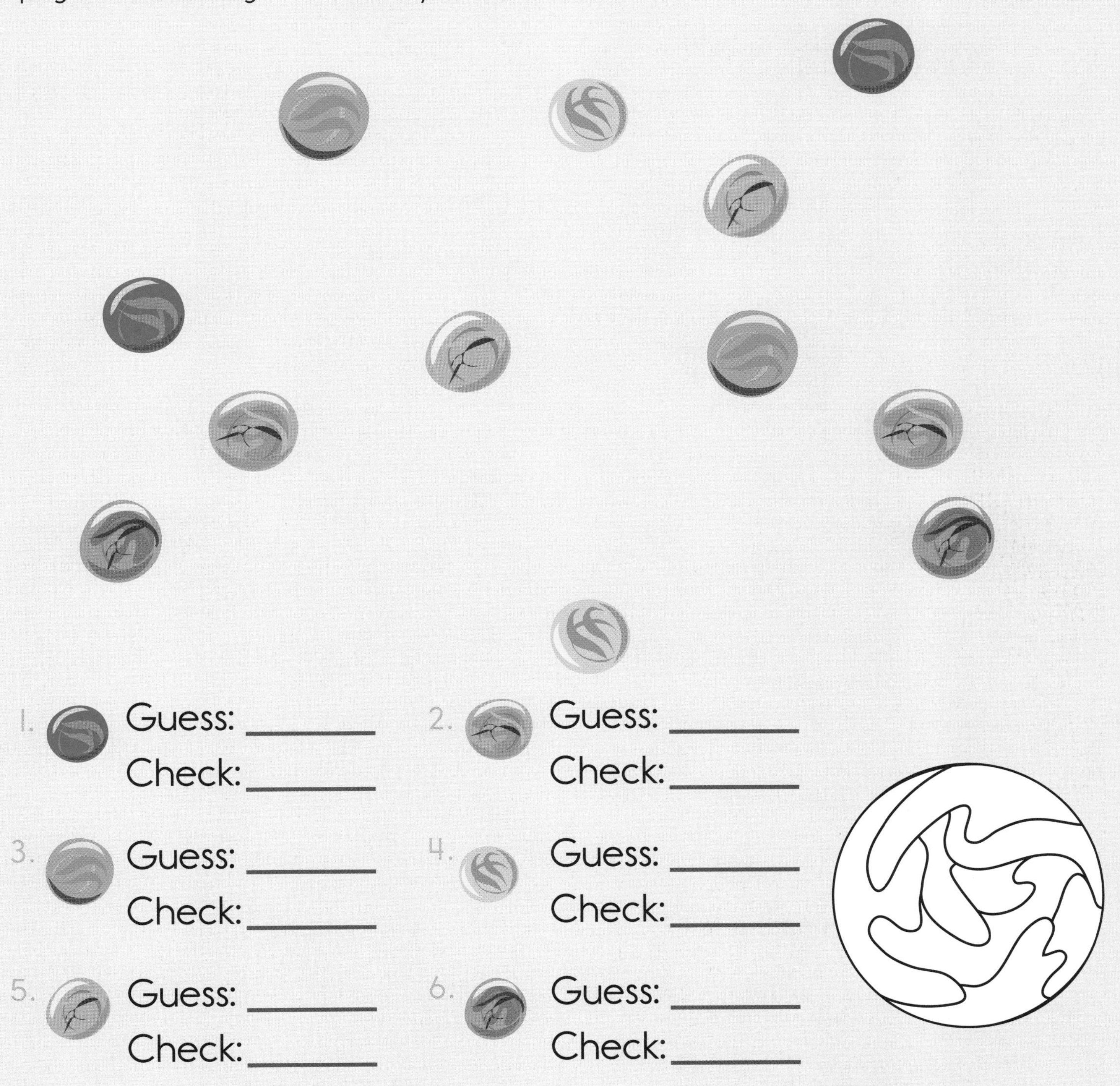

1. Guess: ________
   Check: ________

2. Guess: ________
   Check: ________

3. Guess: ________
   Check: ________

4. Guess: ________
   Check: ________

5. Guess: ________
   Check: ________

6. Guess: ________
   Check: ________

Something's Screwy

One screw is not the same length as the others. MEASURE each screw in centimeters, and CIRCLE the one that is not the same length.

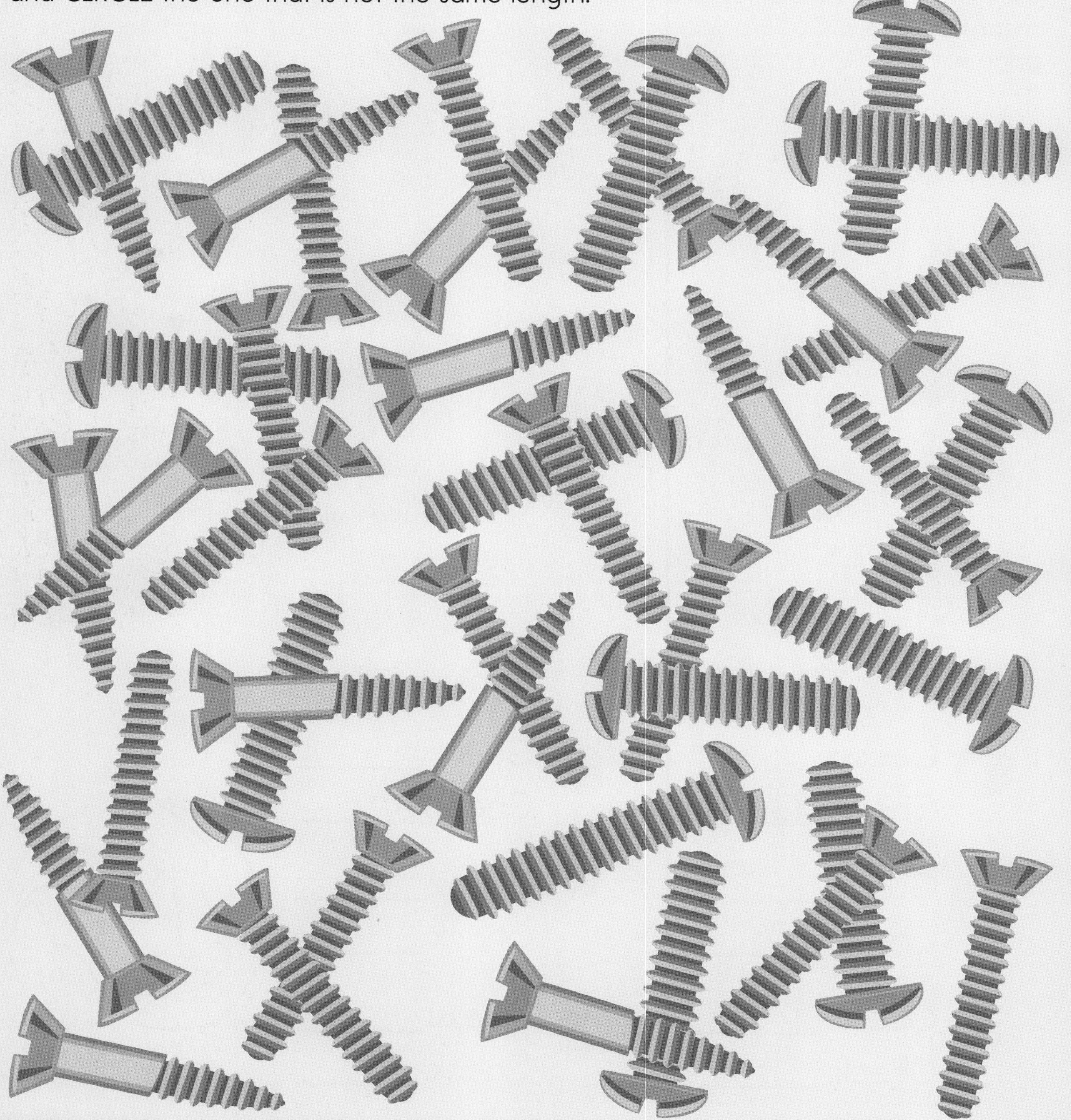

Starry Night

FOLLOW the steps to connect the stars to create the constellation map of Leo the lion. Then GUESS the distance between the star pairs listed, and MEASURE in centimeters to check your guess.

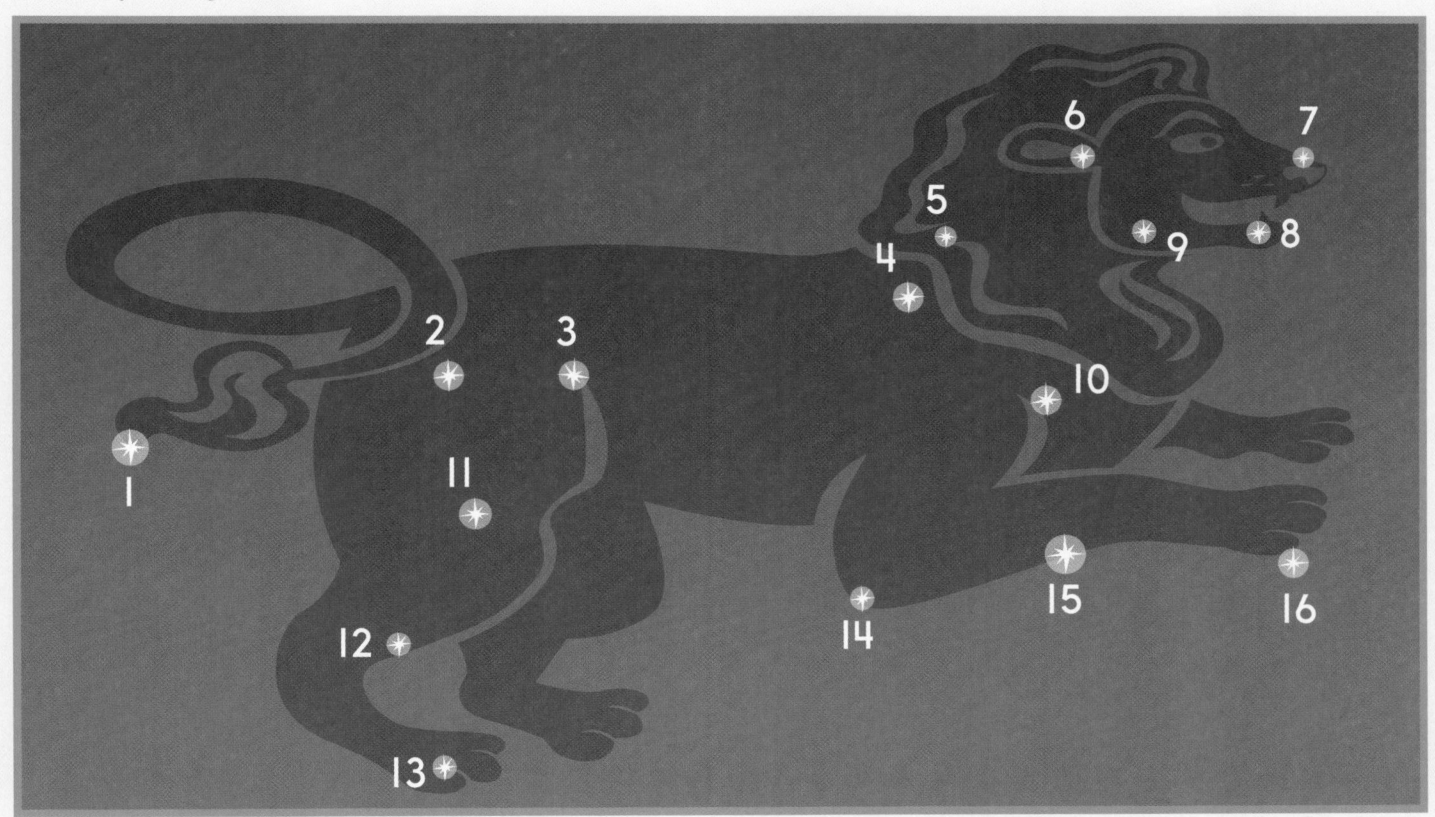

CONNECT stars 1 through 13. Then CONNECT these star pairs:
6 and 9, 4 and 10, 10 and 16, 10 and 15, 2 and 11, 11 and 14.

GUESS the distance between these star pairs, and CHECK your guess.

1. 2 and 11: Guess: ________ cm Check: ________ cm
2. 4 and 6: Guess: ________ cm Check: ________ cm
3. 11 and 15: Guess: ________ cm Check: ________ cm
4. 9 and 16: Guess: ________ cm Check: ________ cm
5. 1 and 14: Guess: ________ cm Check: ________ cm
6. 12 and 15: Guess: ________ cm Check: ________ cm

Code Breaker

WRITE the letter that matches each time to solve the riddle.

O

L

C

K

A

What has two hands but can't carry anything?

____	____	____	____	____	____ .
10:30	6:00	8:30	2:00	6:00	12:30

Mystery Time

COLOR the times in the picture according to the color of the clocks at the top.
When you are done coloring, WRITE the mystery time under the picture.

:

Code Breaker

WRITE the letter that matches each time to solve the riddle.

What happens once in a year, twice in a week, and once in a minute?

___ ___ ___
7:45 3:45 10:15

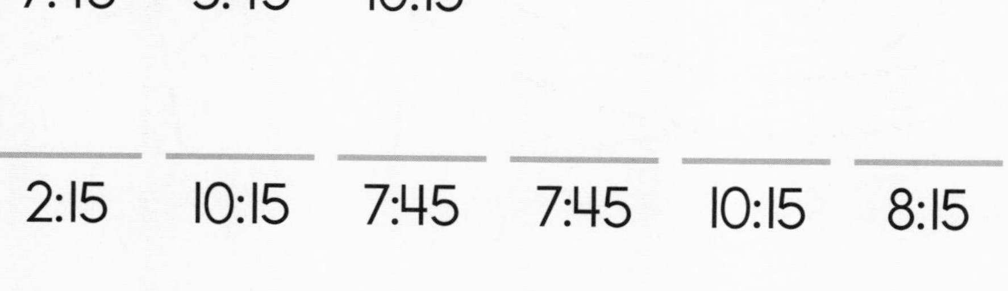

Mystery Time

COLOR the times in the picture according to the color of the clocks at the top.
When you are done coloring, WRITE the mystery time under the picture.

___ : ___

Time Travel

DRAW a line from Start through the clocks to get to the end, traveling
1 hour and 15 minutes ahead as you go from clock to clock.

Start

End

Time Travel

DRAW a line from Start through the clocks to get to the end, traveling 1 hour and 45 minutes back as you go from clock to clock.

Start

End

Code Breaker

WRITE the new time for each watch. Then WRITE the letter that matches each new time to solve the riddle.

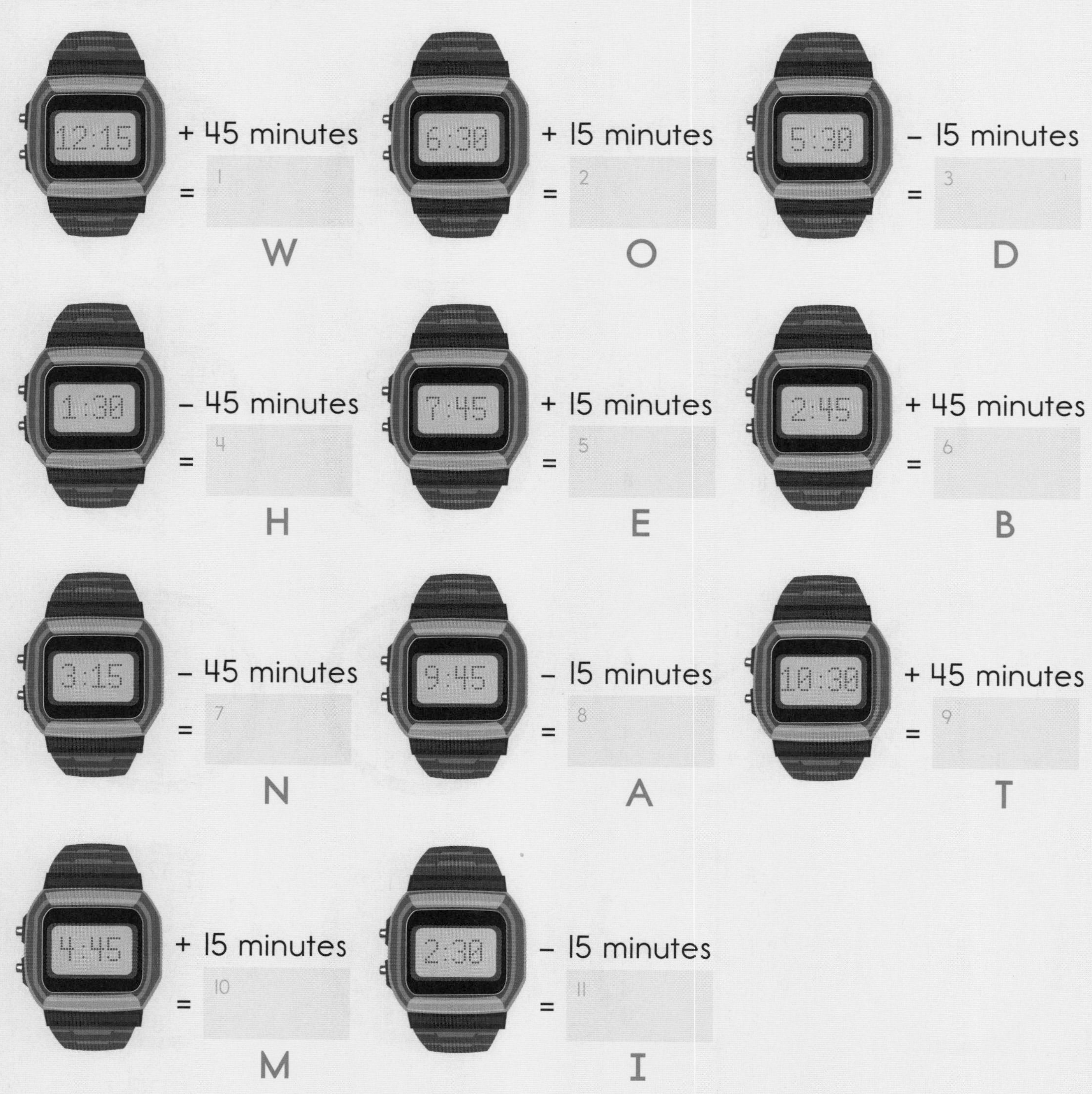

Why did the boy sit on the clock?

____	____		____	____	____	____	____	____
12:45	8:00		1:00	9:30	2:30	11:15	8:00	5:15

____	____		____	____		____	____
11:15	6:45		3:30	8:00		6:45	2:30

____	____	____	____.
11:15	2:15	5:00	8:00

Who Am I?

READ the clues, and CIRCLE the mystery time.

HINT: Cross out any time that does not match the clues.

I am later than 2:00.

I am earlier than 11:00.

I am two hours away from one of the clocks next to me.

In 15 minutes I will be 5:00.

Who am I?

Time Travel

DRAW a line from Start through the clocks to get to the end, traveling 4 hours and 45 minutes back as you go from clock to clock.

Code Breaker

WRITE the value of each coin set. Then WRITE the letter that matches each value to solve the riddle.

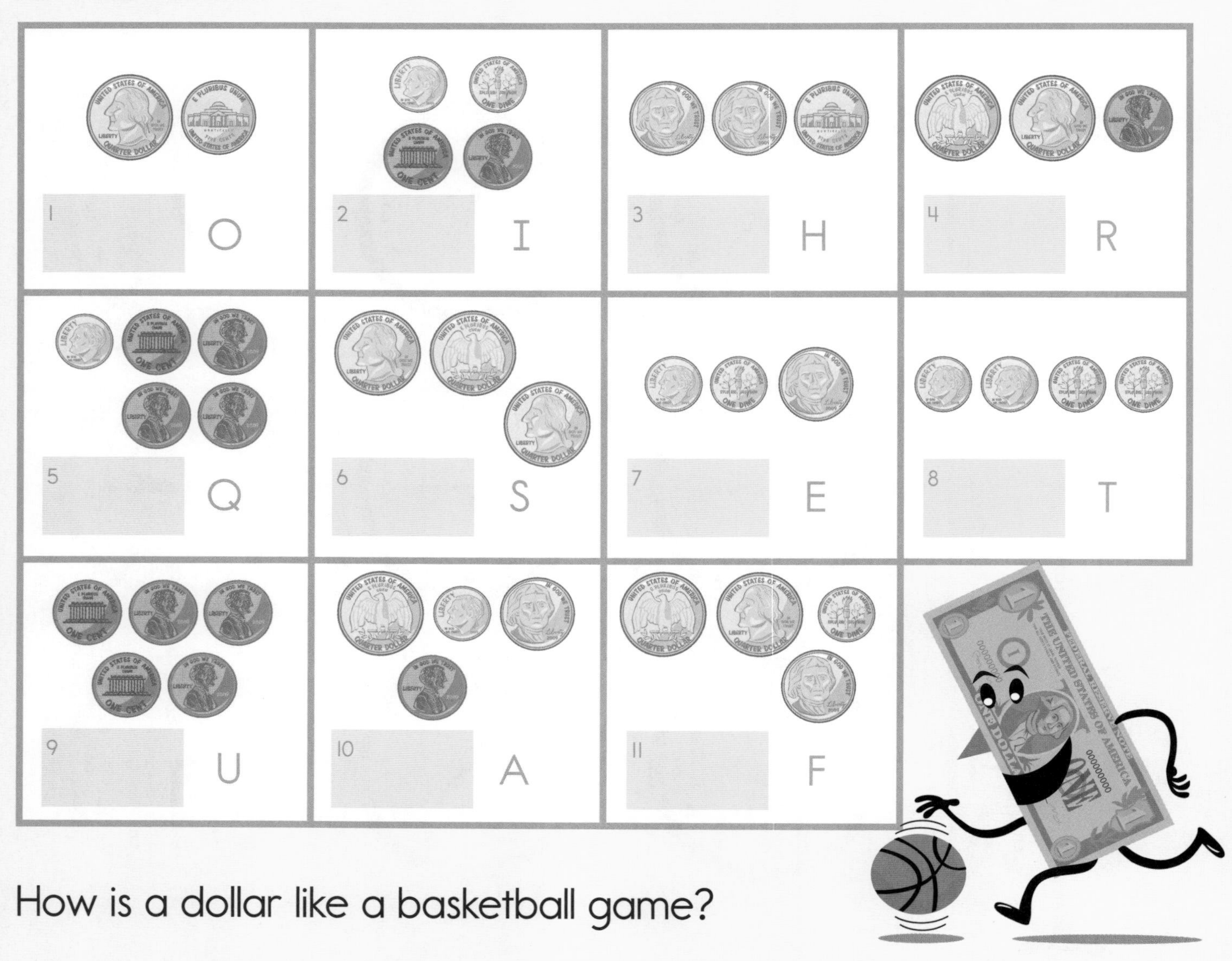

How is a dollar like a basketball game?

22¢ 40¢ 15¢ 41¢ 75¢ 65¢ 30¢ 5¢ 51¢

14¢ 5¢ 41¢ 51¢ 40¢ 25¢ 51¢ 75¢.

Slide Sort

CIRCLE the dollar amounts that do **not** match the picture at the bottom of the slide.

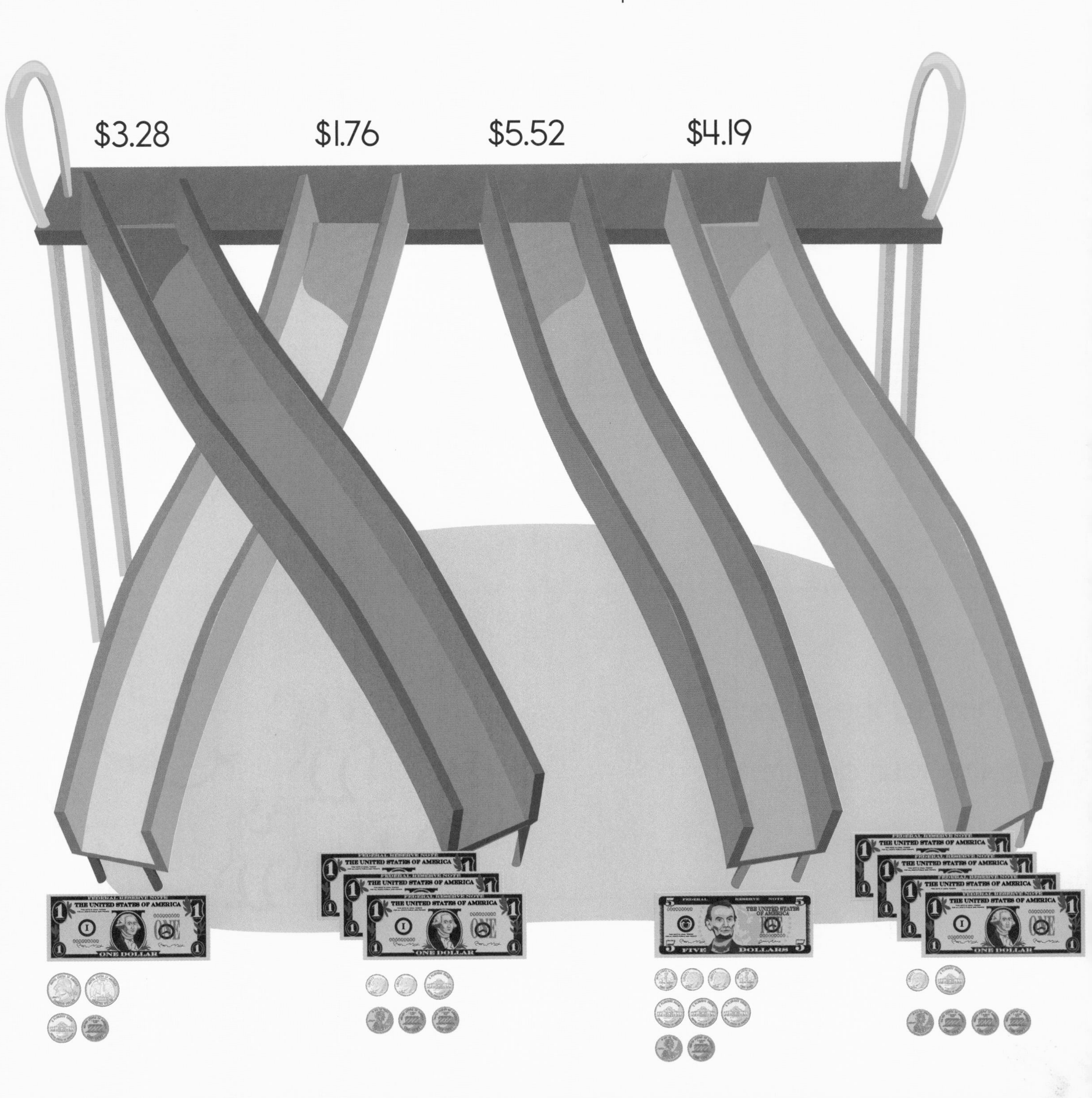

Code Breaker

WRITE the value of each money set. Then WRITE the letter that matches each value to solve the riddle.

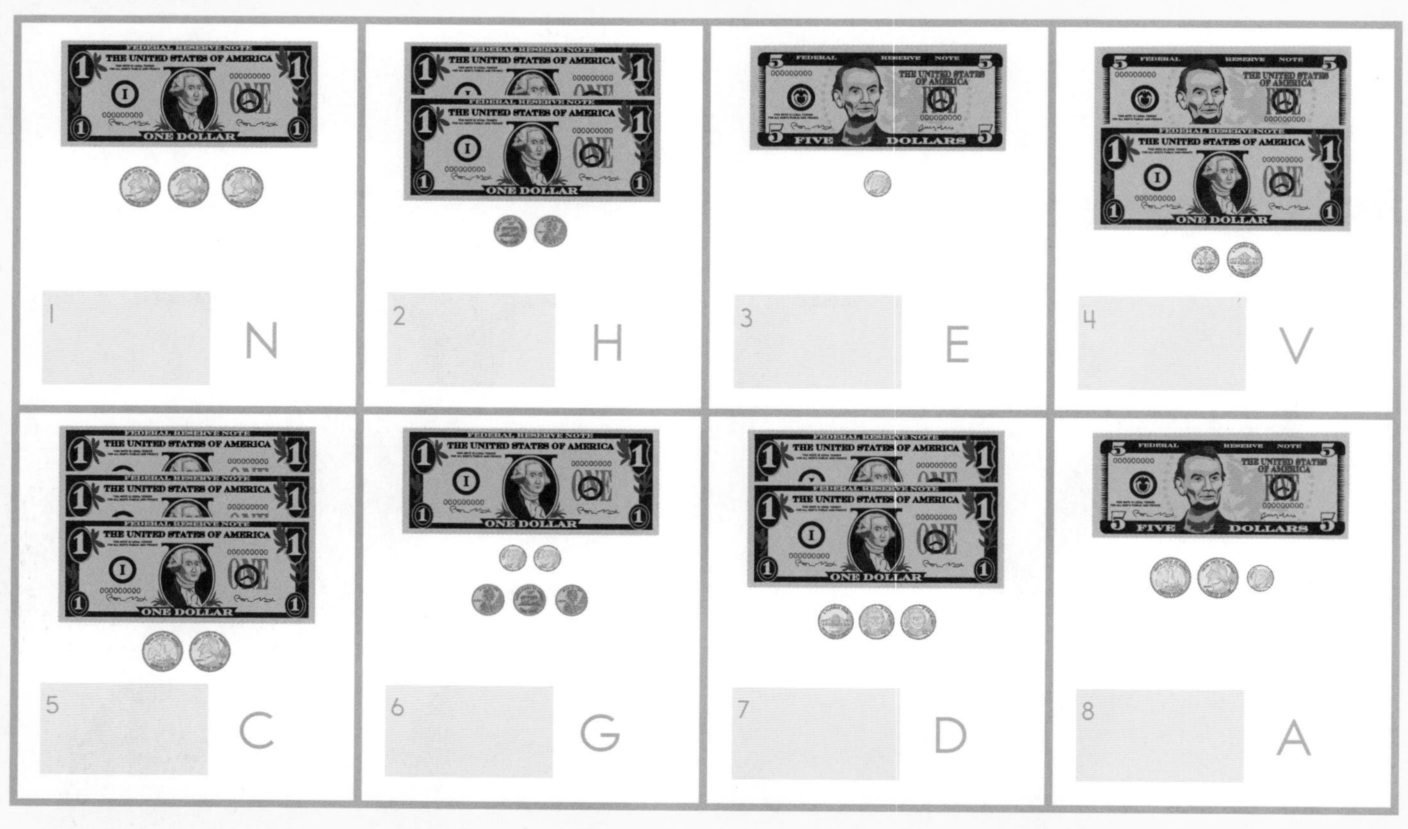

What did the dollar say to the four quarters?

You ______ ______ ______ ______
$2.02 $5.60 $6.15 $5.10

______ ______ ______ ______ ______ ______ ______.
$3.50 $2.02 $5.60 $1.75 $1.23 $5.10 $2.15

Pocket Change

DRAW two straight lines to create four different money sets of equal value.

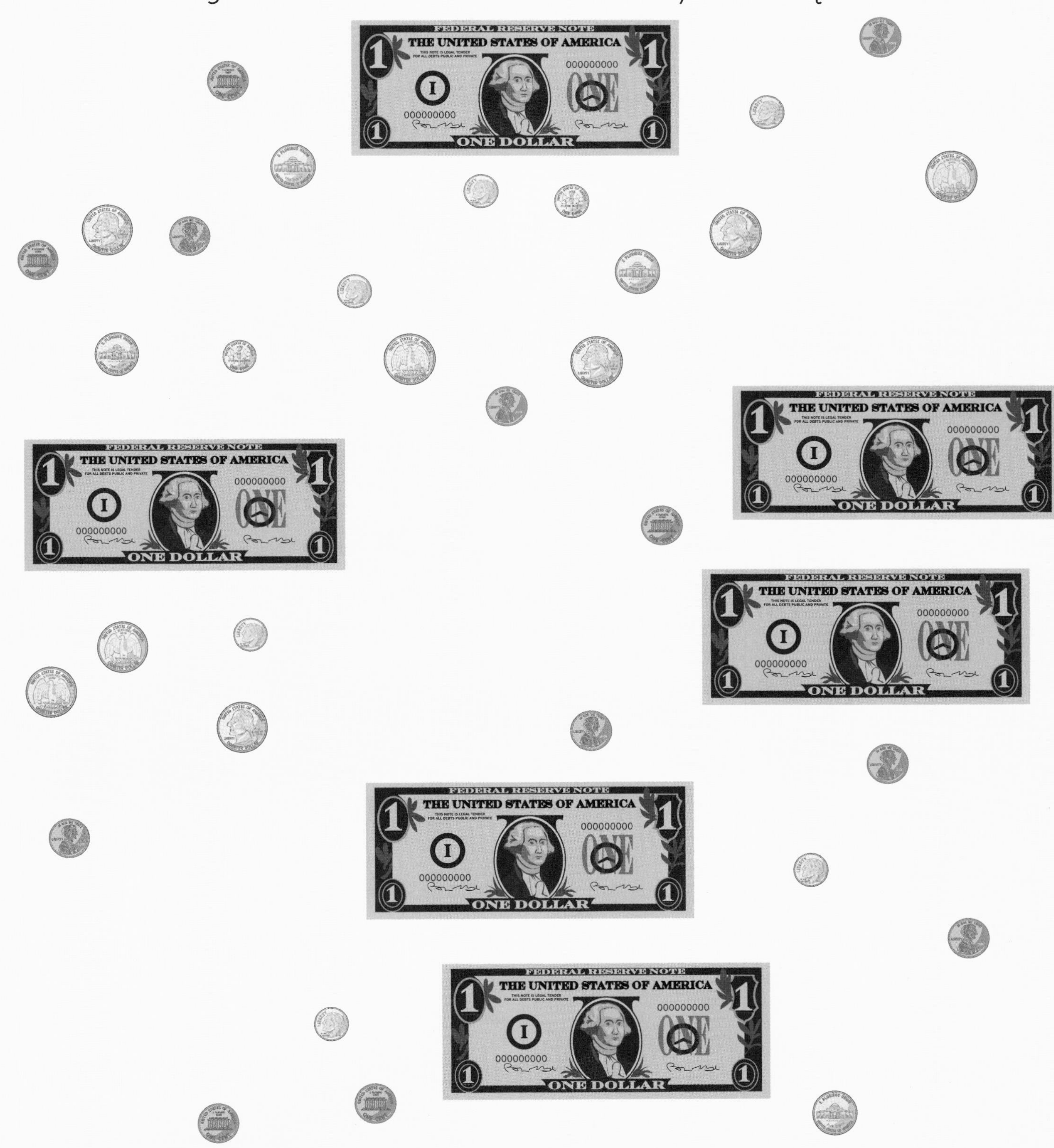

Pay the Price

CUT OUT the cards on pages 103 and 104. PLACE cards next to each price tag so that the cards total the same value. How many different ways can you place the cards for each price tag?

HINT: Use as many cards as you want, stacking them next to each price tag.

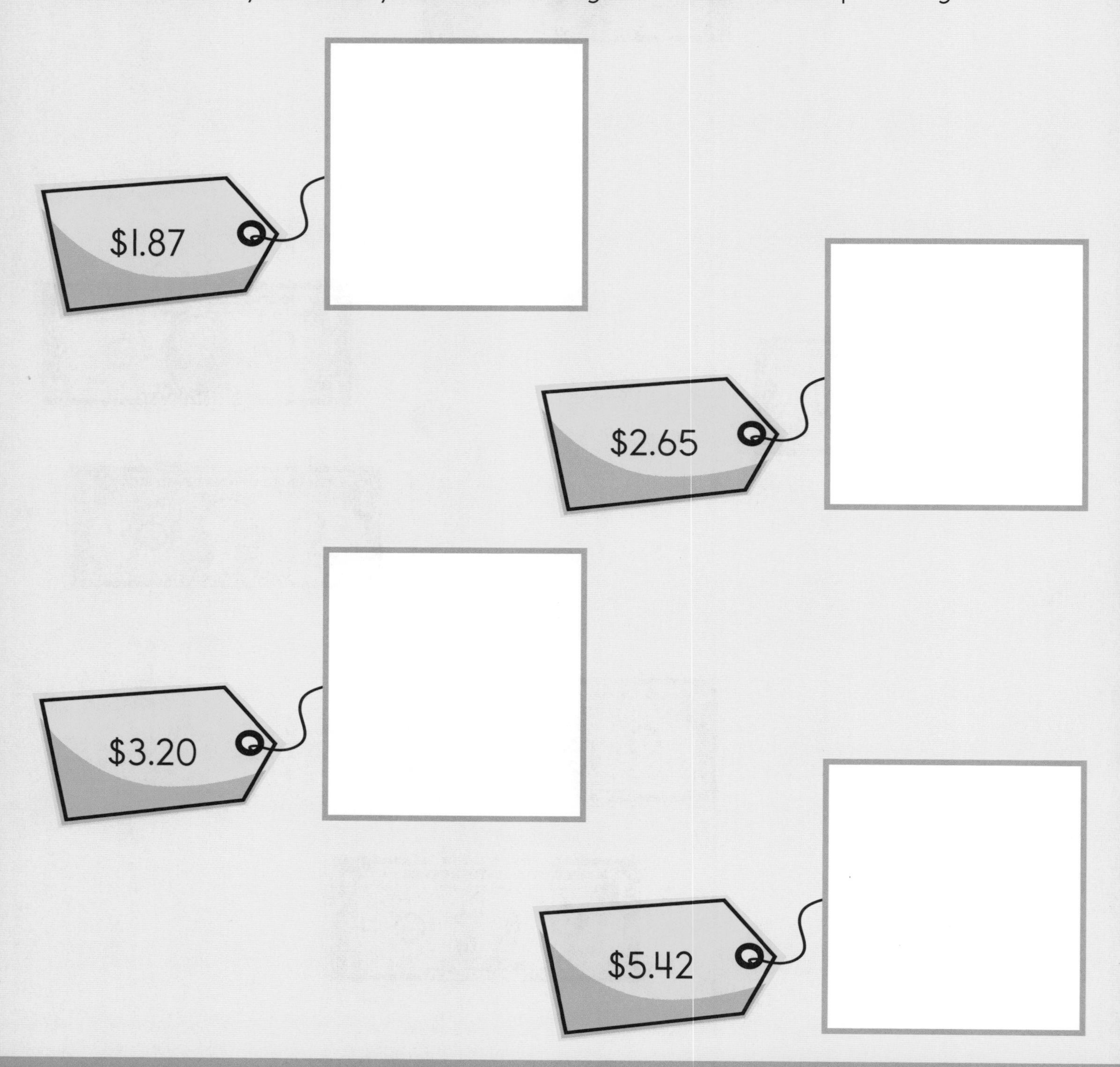

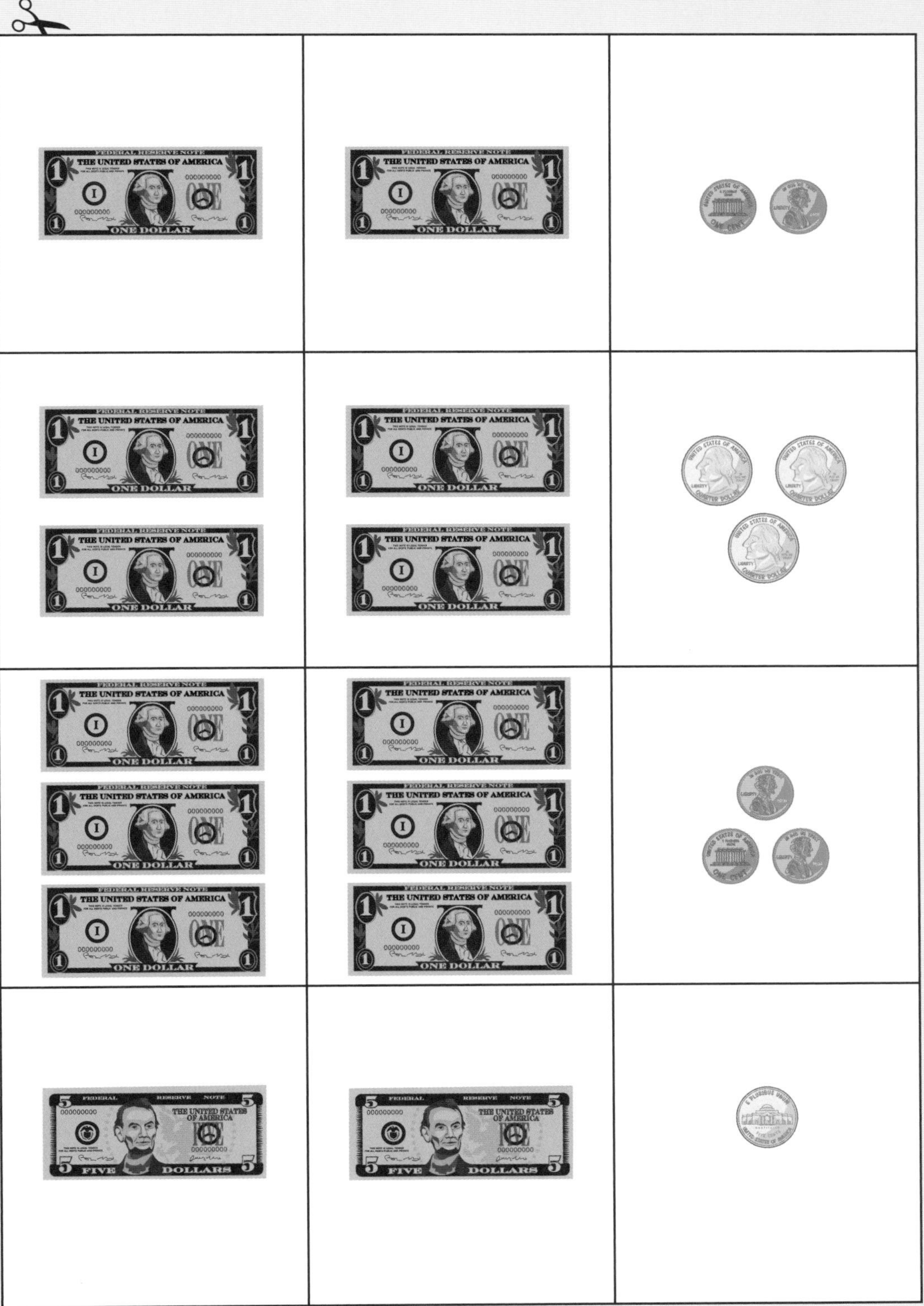
FEDERAL RESERVE NOTE
THE UNITED STATES OF AMERICA
ONE DOLLAR
FEDERAL RESERVE NOTE
THE UNITED STATES OF AMERICA
FIVE DOLLARS

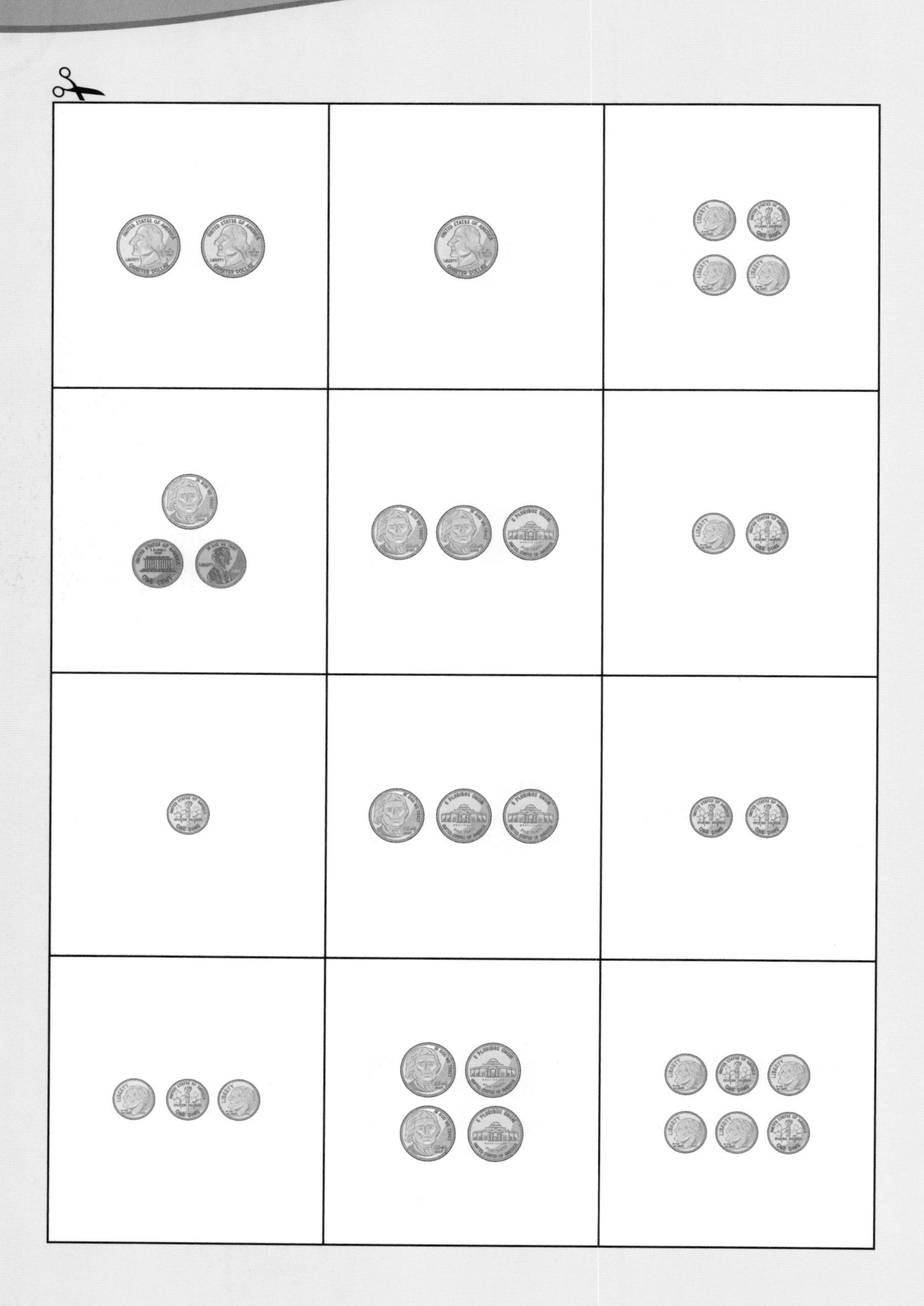

One Left Over

CROSS OUT all of the money that will be used to buy the items to reveal the extra coin.

Code Breaker

CIRCLE the picture in each row with less money than the other, and WRITE its value. Then WRITE the letter that matches each value to solve the riddle.

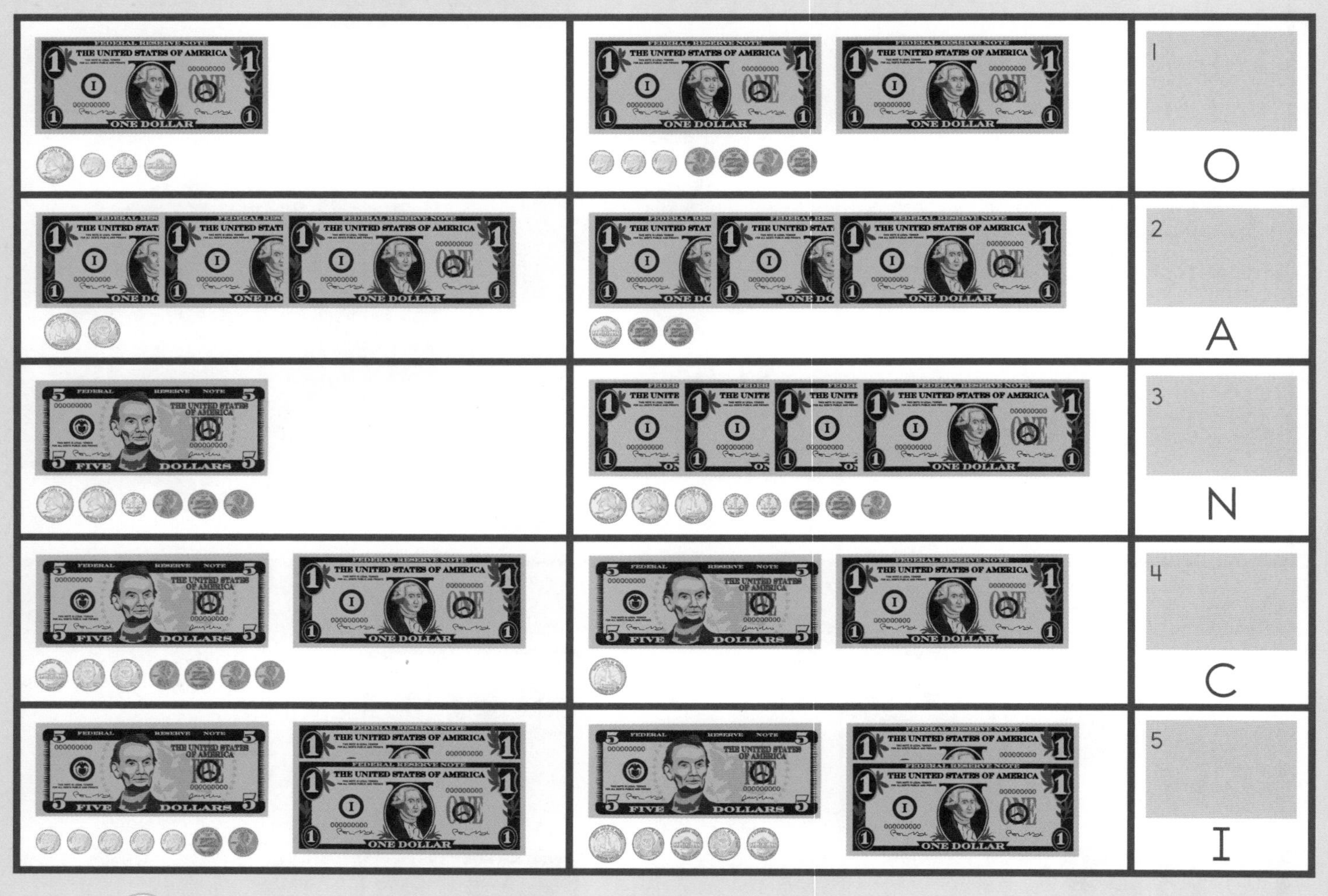

What has a head and a tail but no body?

_____ _____ _____ _____ _____.

$3.07 $6.19 $1.50 $7.45 $4.98

Just Right

WRITE these dollar amounts so that each one is next to a picture with a smaller value.

HINT: There may be more than one place to put each amount, but you need to use every one.

$2.29 $4.98 $5.14 $1.03 $6.37

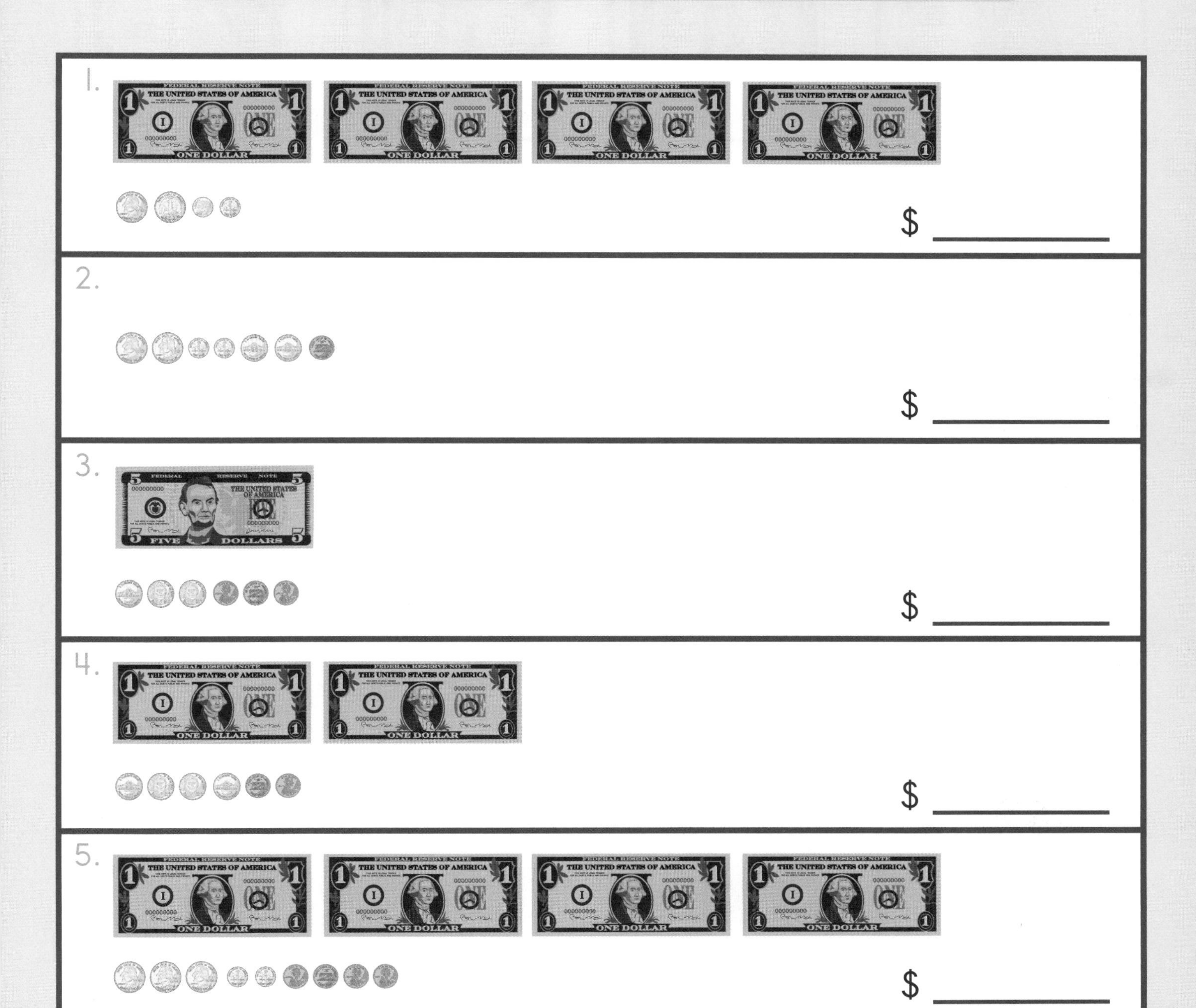

Code Breaker

WRITE the value of each money set. Then WRITE the letter that matches each value to solve the riddle.

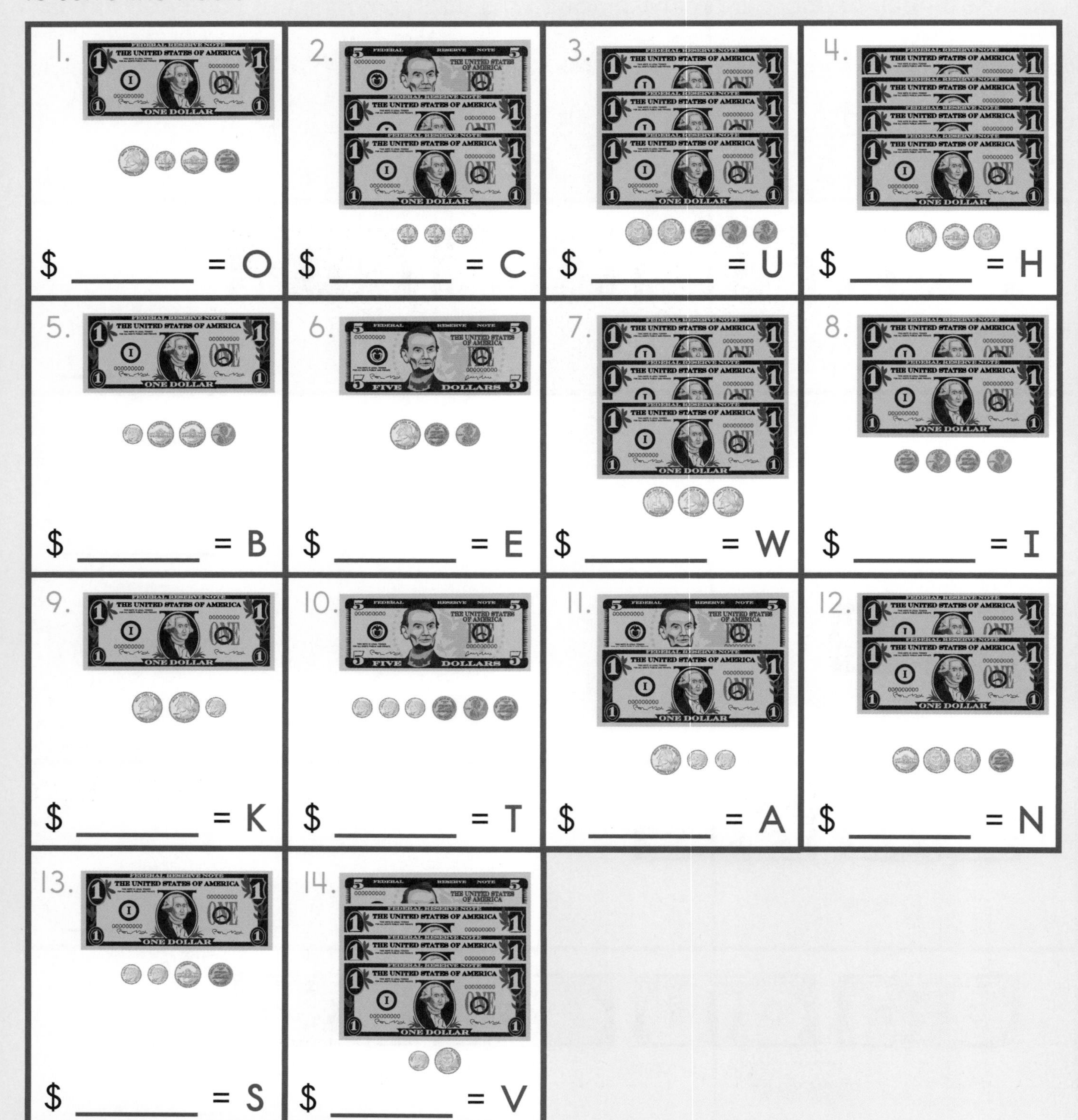

Why did the elephant cross the road?

\$1.21 \$5.27 \$7.30 \$6.45 \$3.13 \$1.26 \$5.27

\$5.33 \$4.35 \$5.27   \$7.30 \$4.35 \$2.04 \$7.30 \$1.60 \$5.27 \$2.16

\$3.75 \$6.45 \$1.26   \$1.41 \$2.16

\$8.15 \$6.45   \$7.30 \$6.45 \$5.33 \$2.04 \$1.41 \$2.16 .

Pocket Change

DRAW three straight lines to create six different money sets of equal value.

Beans

CUT OUT the beans.

These beans are for use with pages 40, 42, and 43.

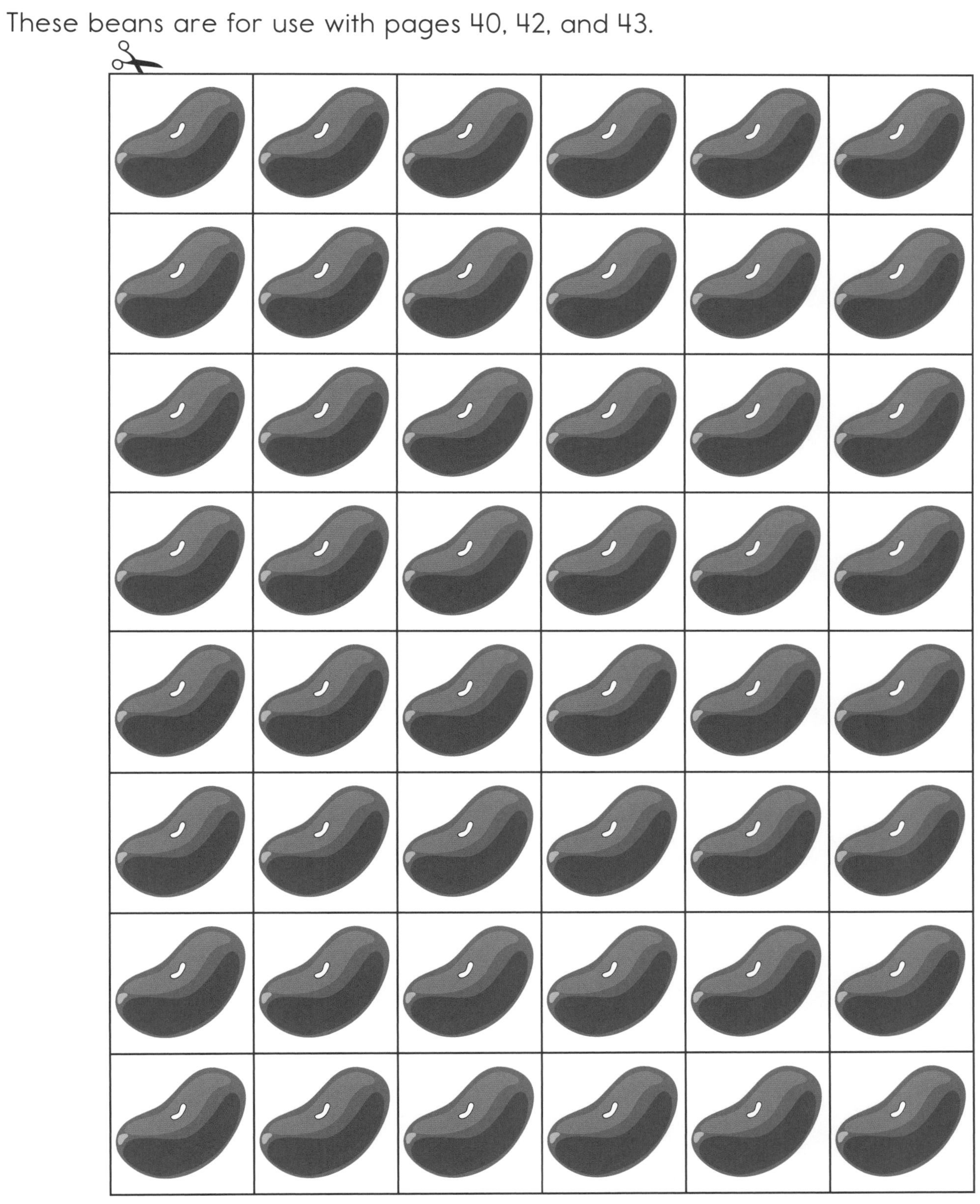

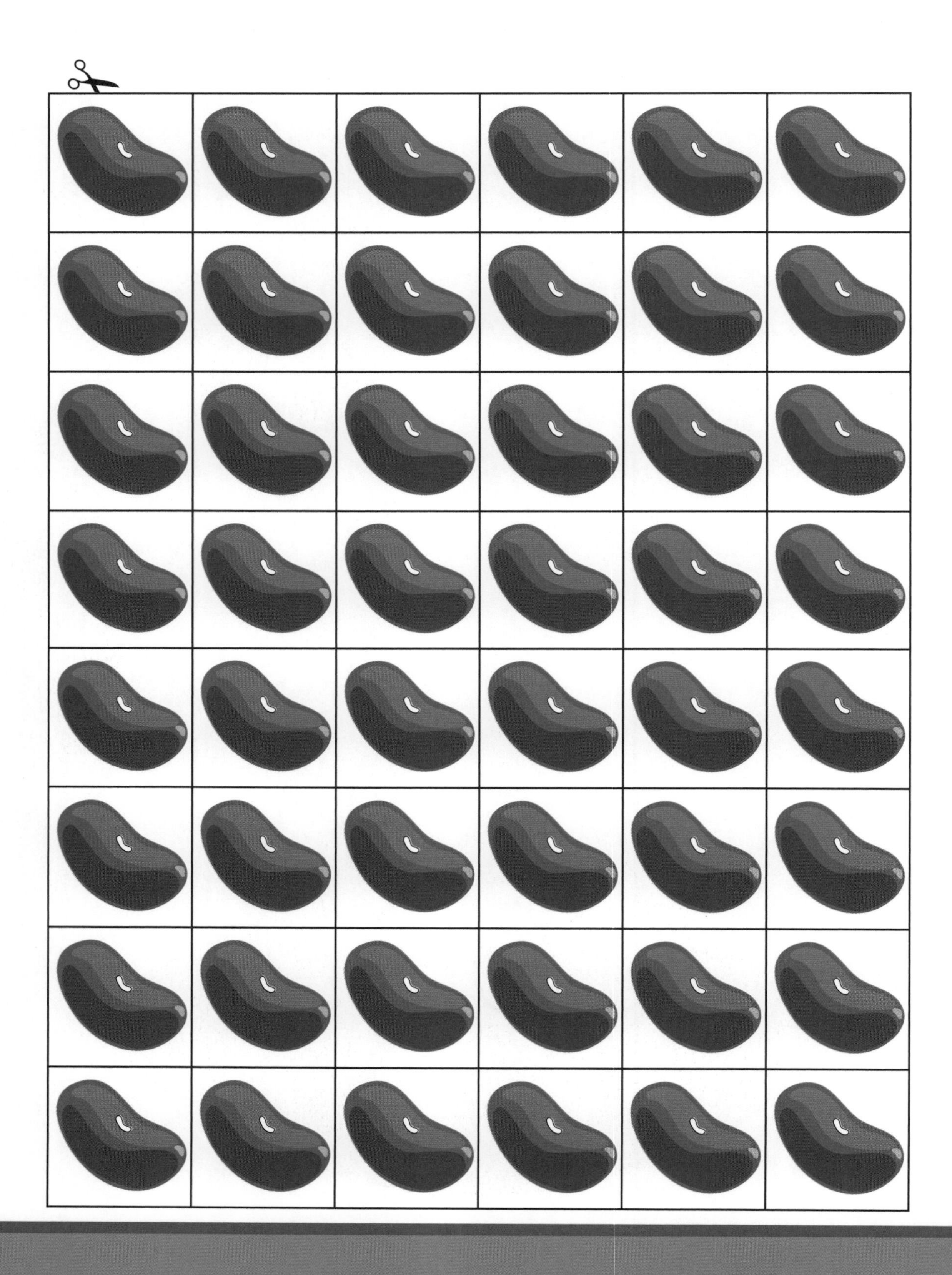

Spinners

CUT OUT the spinner. BEND the outer part of a paper clip so that it points out, and carefully POKE it through the center dot of the spinner. You're ready to spin!

This spinner is for use with page 47, and the reverse side is for use with pages 54 and 55.

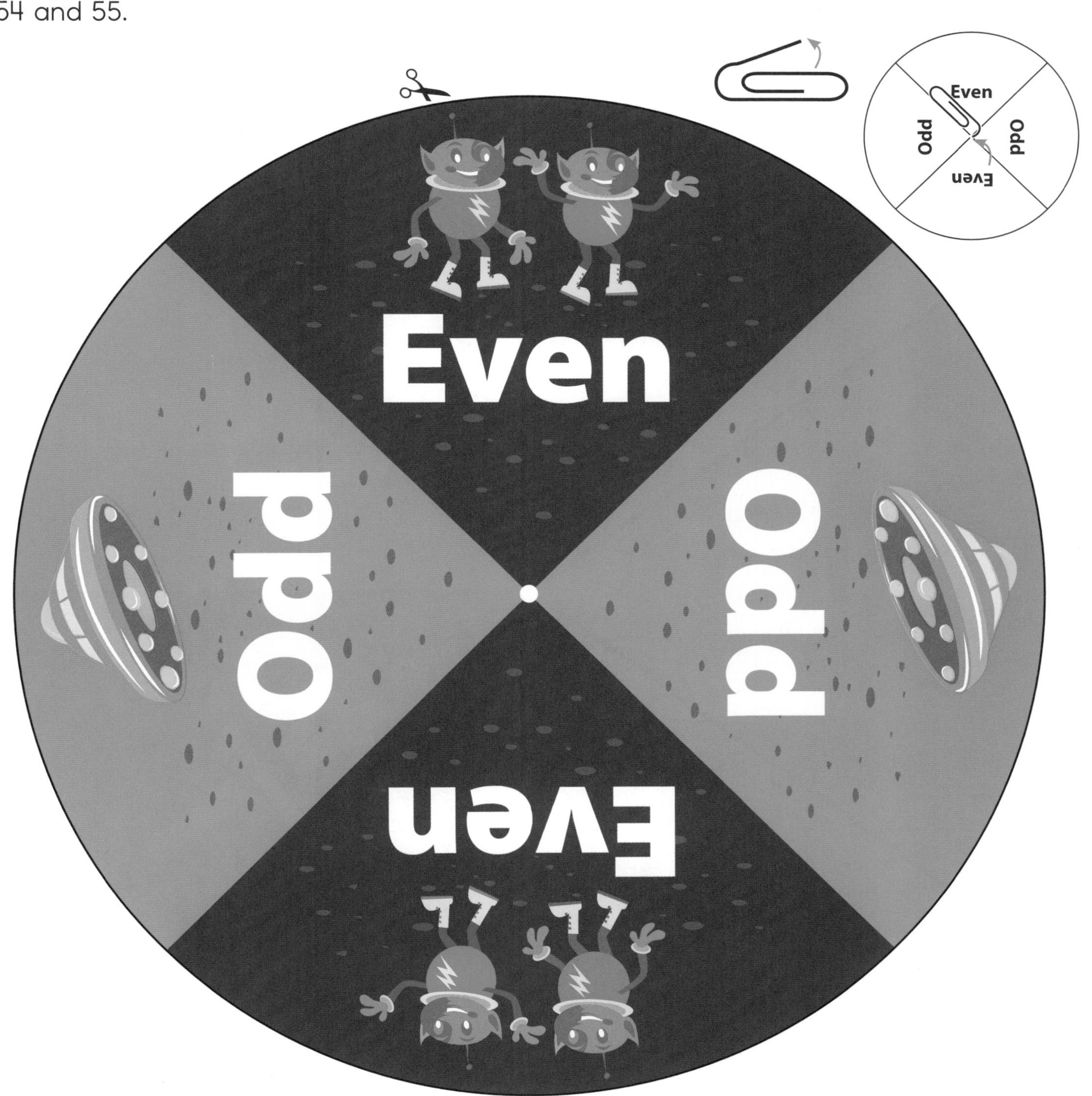

Use the spinner on this side for pages 54 and 55. Pull out the paper clip from the other side, and poke it through the center dot on this side.

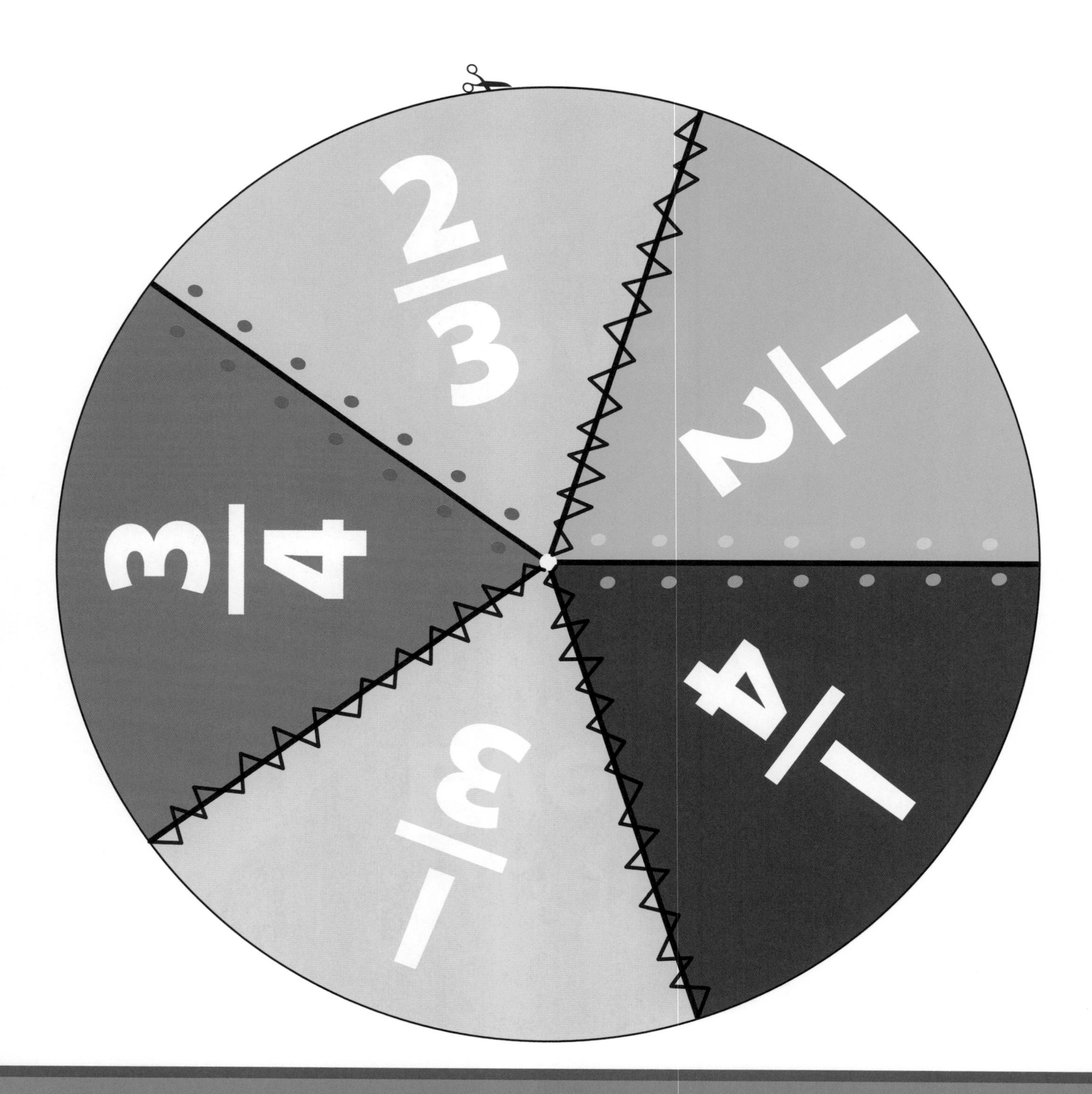

Pattern Blocks

CUT OUT the 31 pattern block pieces.

These pattern block pieces are for use with pages 64, 66, 67, and 74.

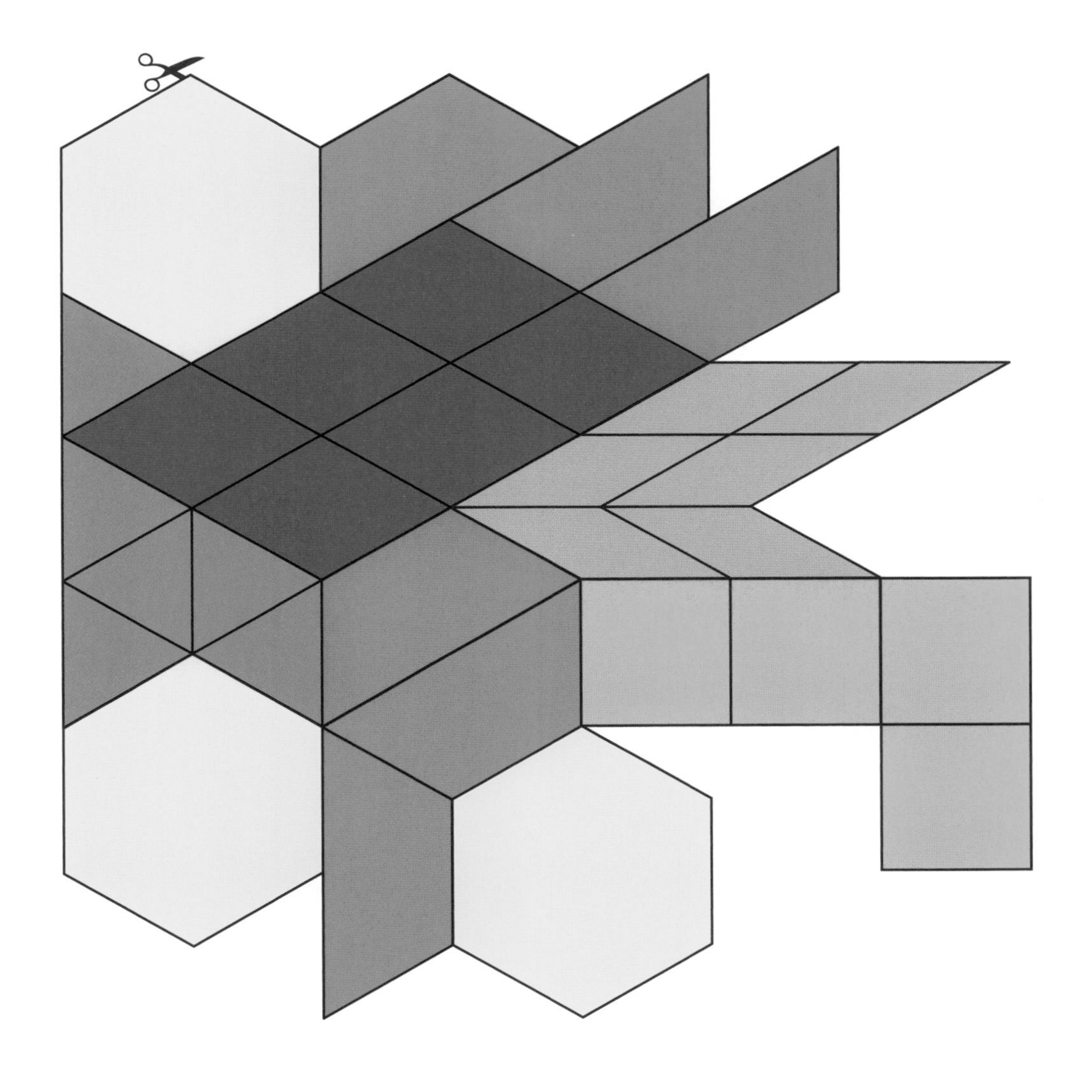

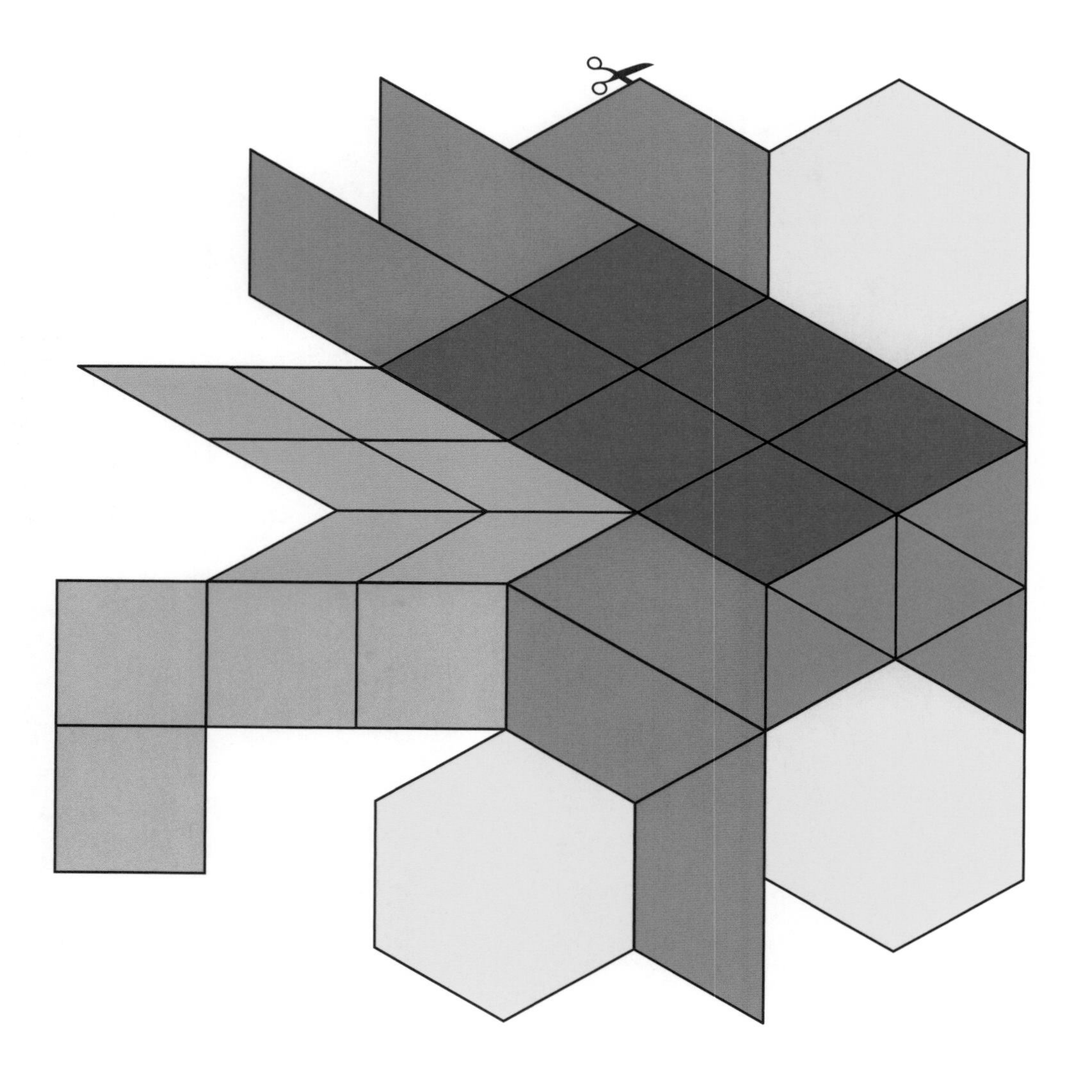

Pentominoes

CUT OUT the 13 pentomino pieces.

These pentomino pieces are for use with pages 68, 70, and 75.

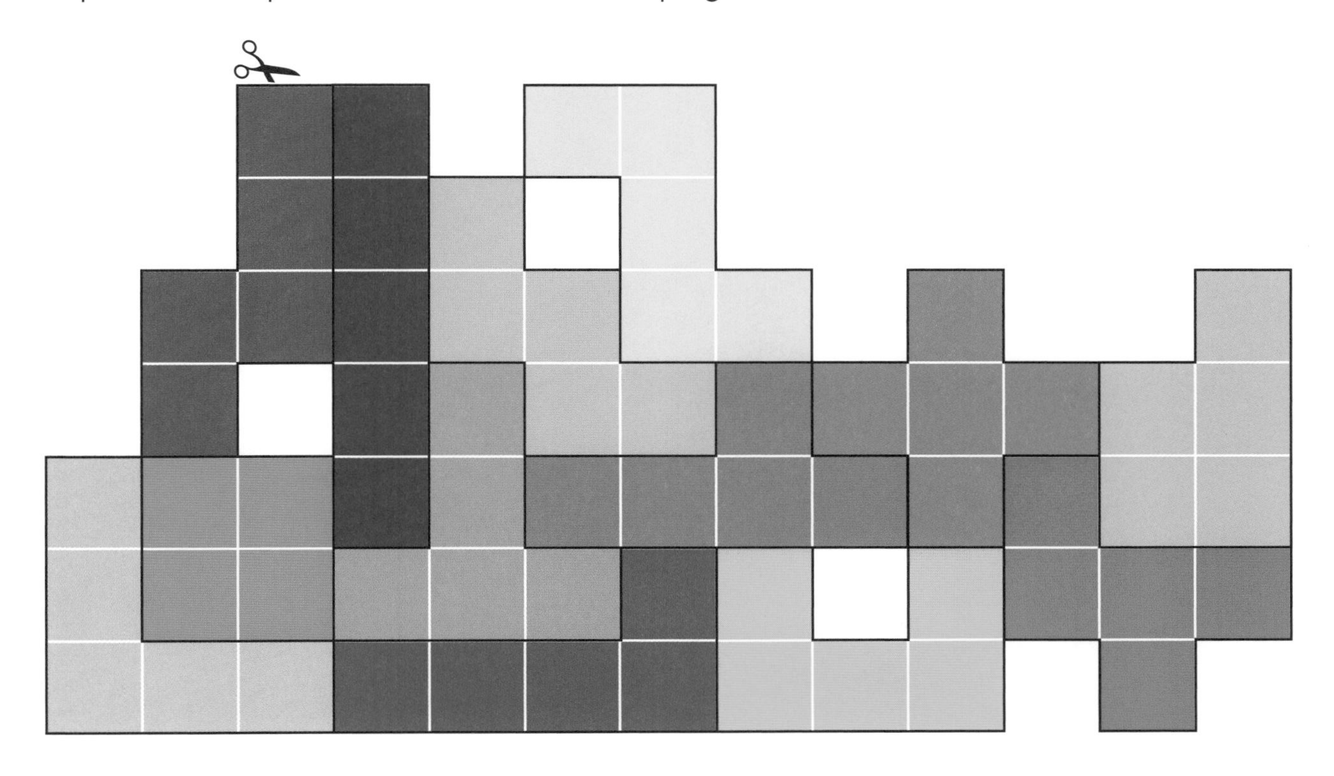

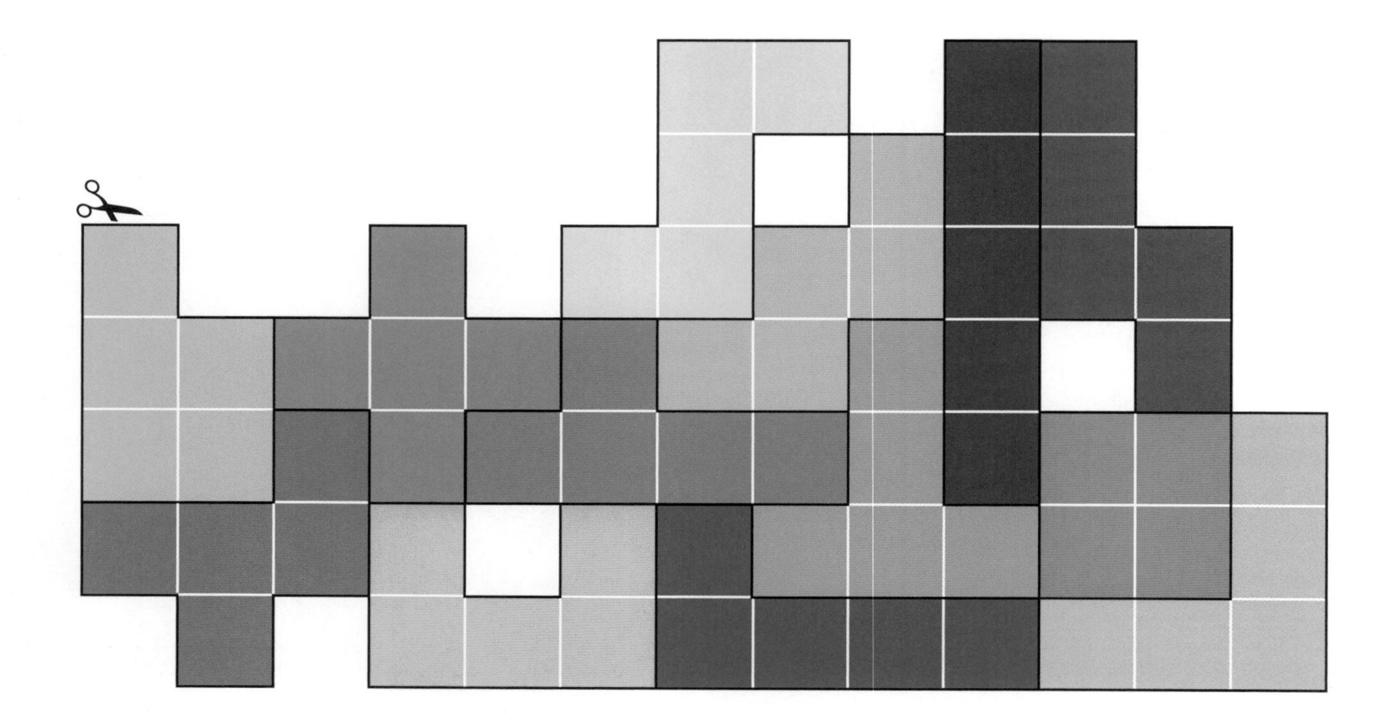

Page 2

313	213	[illegible]	321	[illegible]	213	313	321
213	[illegible]	321	313	321	[illegible]	213	313
[illegible]	321	313	213	313	321	[illegible]	213
321	313	213	[illegible]	213	313	321	[illegible]
313	213	[illegible]	321	[illegible]	213	313	321
213	[illegible]	321	313	321	[illegible]	213	313
[illegible]	321	313	213	313	321	[illegible]	213
321	313	213	[illegible]	213	313	321	[illegible]

Page 3

1. 364 2. 416
3. 195 4. 608
5. 241

Combination 3 4 1 6 2

Page 4

1. 159 2. 344
3. 506 4. 830
5. 718 6. 472

5	8	0	3	1	4
2	1	5	4	7	0
0	5	0	6	8	3
8	9	3	1	3	5
3	3	7	6	0	0
1	5	2	3	4	8
7	1	8	4	2	1
1	9	6	4	7	2

Page 5

Have someone check your answers.

Page 6

1. 97, 100, 103
2. 215, 218, 220
3. 746, 747, 752, 753

I LIKE YOUR BELT!

Page 7

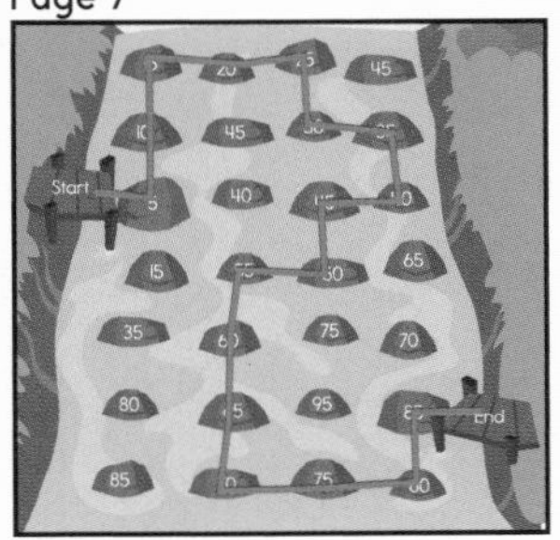

Page 8

1	2	3	4	5	6	7	8	9	10
11	12	13	14	15	16	17	18	19	20
21	22	23	24	25	26	27	28	29	30
31	32	33	34	35	36	37	38	39	40
41	42	43	44	45	46	47	48	49	50
51	52	53	54	55	56	57	58	59	60
61	62	63	64	65	66	67	68	69	70
71	72	73	74	75	76	77	78	79	80
81	82	83	84	85	86	87	88	89	90
91	92	93	94	95	96	97	98	99	100

Page 9

10	4	7		175	116	330
20	8	14		168	112	320
30	12	21		161	108	310
40	16	28		154	104	300
50	20	35		147	100	290
60	24	42		140	96	280
70	28	49		133	92	270
80	32	56		126	88	260
90	36	63		119	84	250
100	40	70		112	80	240
110	44	77		105	76	230
120	48	84	91	98	72	220
130	52	56	60	64	68	210
140	150	160	170	180	190	200

Page 10

1. 677 2. 113
3. 742 4. 556
5. 823 6. 256
7. 981 8. 409
9. 187 10. 399

Page 11

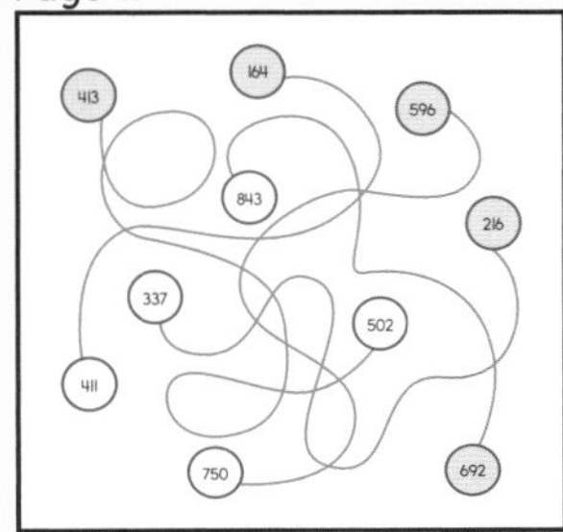

Page 12

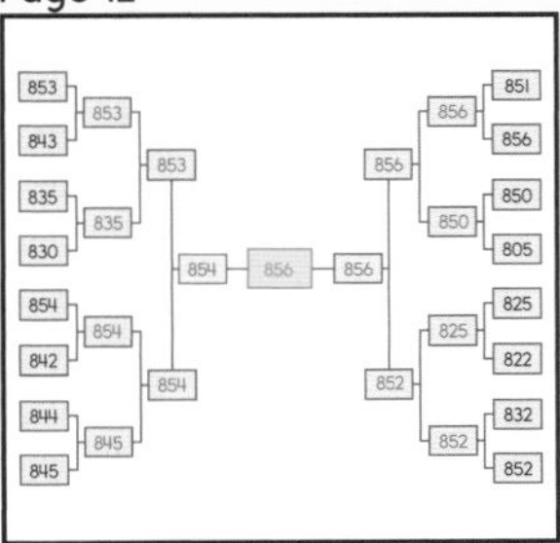

Page 13

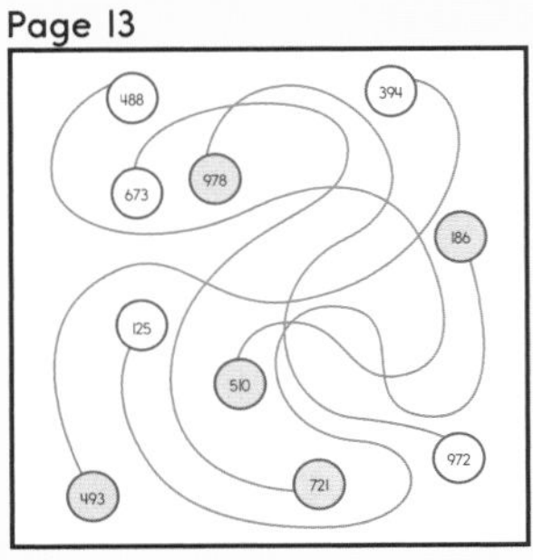

Page 14

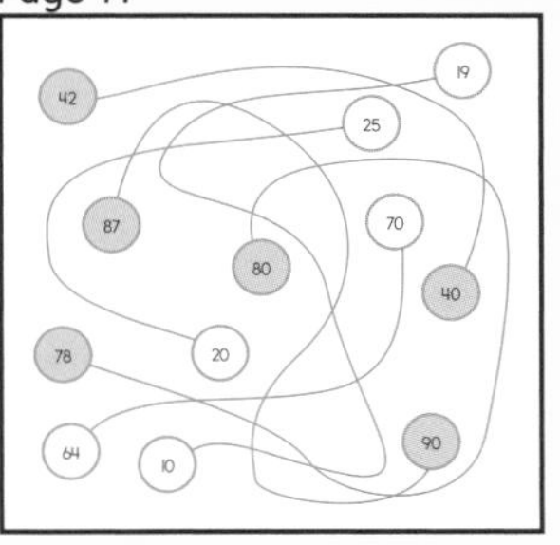

Page 15

Have someone check your answers.

Page 16

1. 285 2. 544
3. 709 4. 952
5. 549 6. 278
7. 932 8. 717
9. 751

Pages 17–18

Check: 117

Page 19

Check: 81

Page 20

6	12	18	24	30	36	42
144	150	156	162	168	174	48
138	240	246	252	258	180	54
132	234	288	294	264	186	60
126	228	282	276	270	192	66
120	222	216	210	204	198	72
114	108	102	96	90	84	78

Page 21

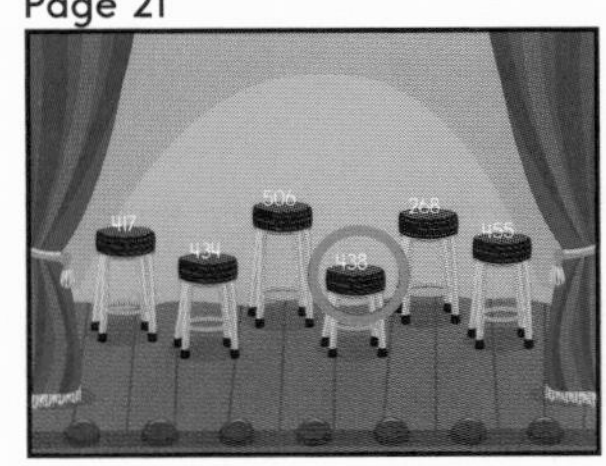

Page 22

1. 21 2. 16
3. 45 4. 37
5. 62 6. 54

Page 23

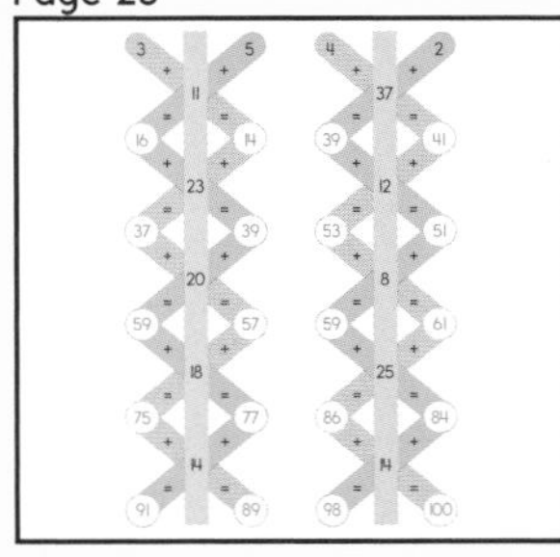

Page 24

3	+	12	=	15
+		+		+
9	+	25	=	34
=		=		=
12	+	37	=	49

Page 25

1. 37 2. 68
3. 59 4. 79
5. 95 6. 48
7. 93 8. 77
9. 63 10. 71
11. 80 12. 91

IT GOES TO PENCILVANIA.

Page 26

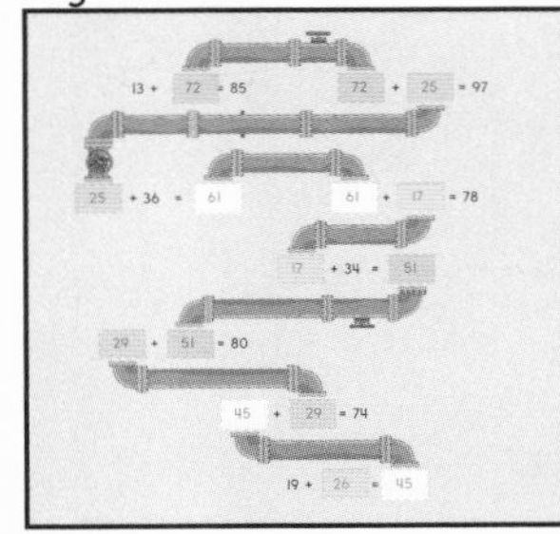

Page 27

Suggestion:

Page 28

1. 12 2. 33
3. 71 4. 48
5. 22 6. 56

Answers

Page 29

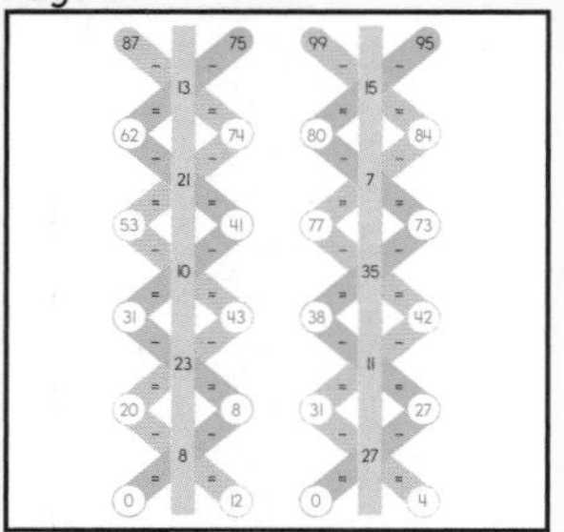

Page 30

95	-	63	=	32
-		-		-
47	-	17	=	30
=		=		=
48	-	46	=	2

Page 31

1. 31 2. 23
3. 2 4. 22
5. 46 6. 49
7. 26 8. 52
9. 13 10. 47
11. 36 12. 16

YOU WOULD HAVE TWO TOYS.

Page 32

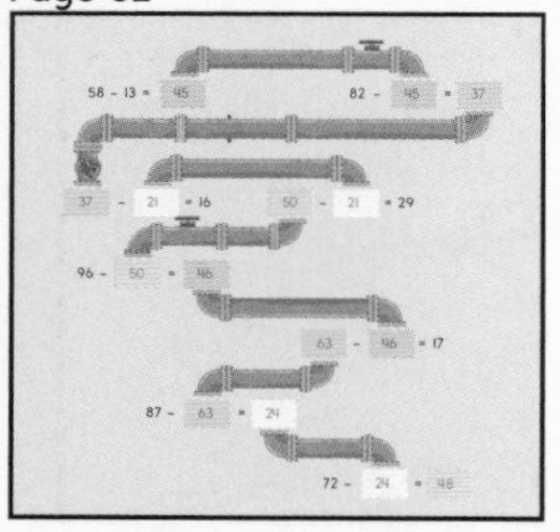

Page 33
Suggestion:

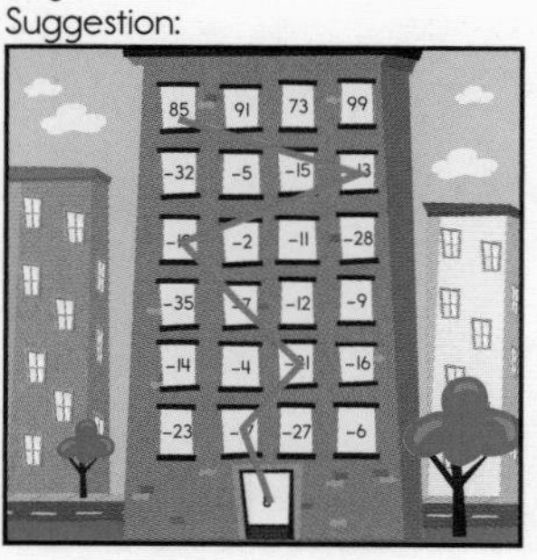

Page 34

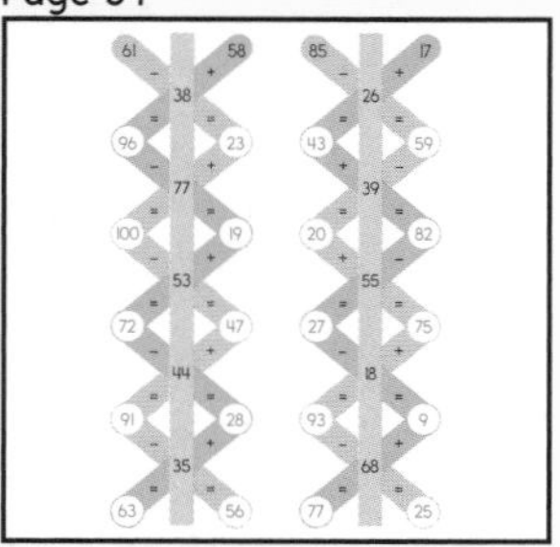

Page 35
Suggestion:

Page 36

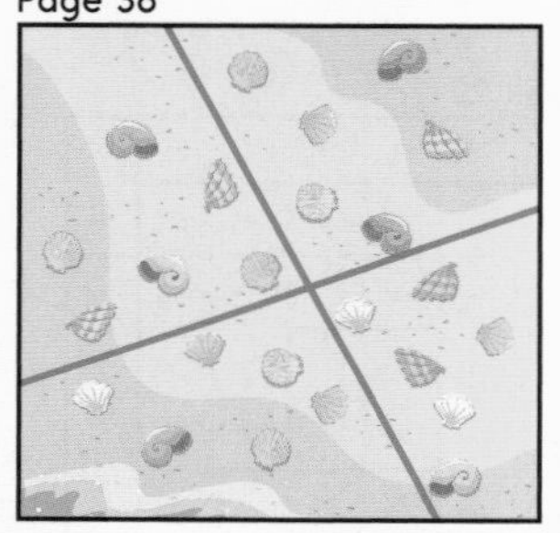

Page 37

1. 3 2. 2
3. 6 4. 2
5. 1 6. 3

Page 38
Suggestion:

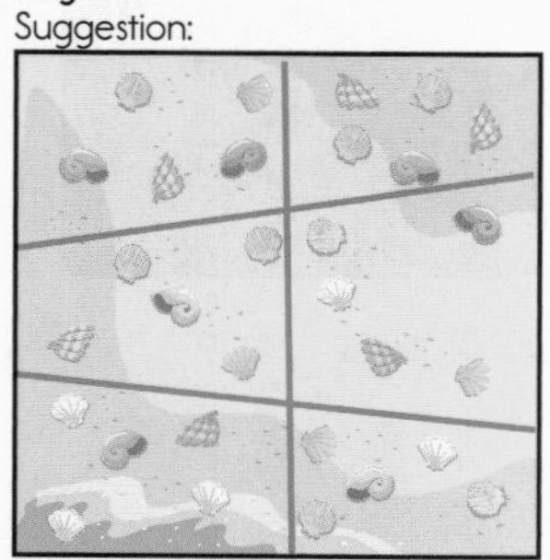

Page 39

1. 2 2. 4
3. 2 4. 5
5. 4 6. 2

Page 40

1. 3 2. 5
3. 2 4. 6
5. 8 6. 4

Page 41
Suggestion:

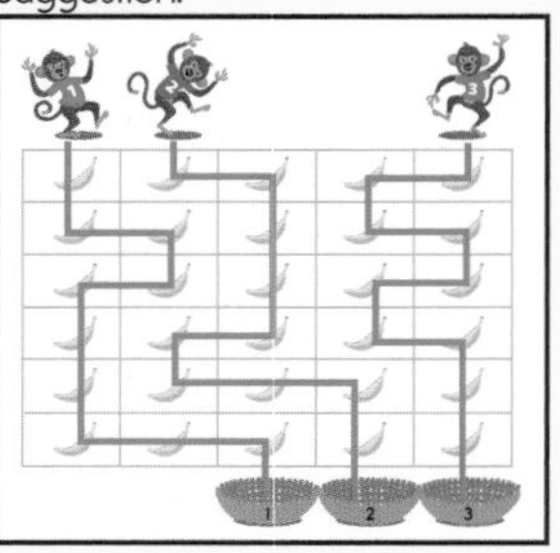

Page 42

1. 4 2. 8
3. 2 4. 7
5. 5 6. 12

Page 44

1. 3 2. 7
3. 9 4. 1
5. 6 6. 12
7. 8 8. 10

TAKE AWAY THE S.

Page 45

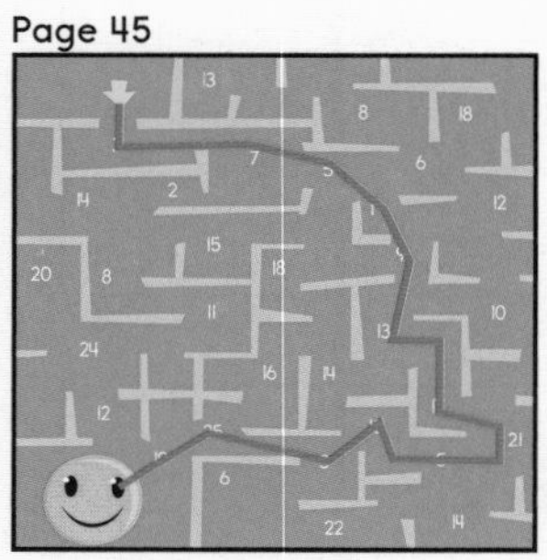

Page 46

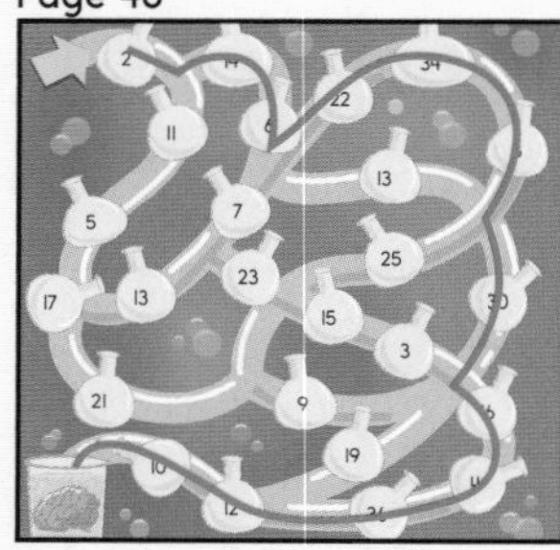

Page 48
Suggestion:

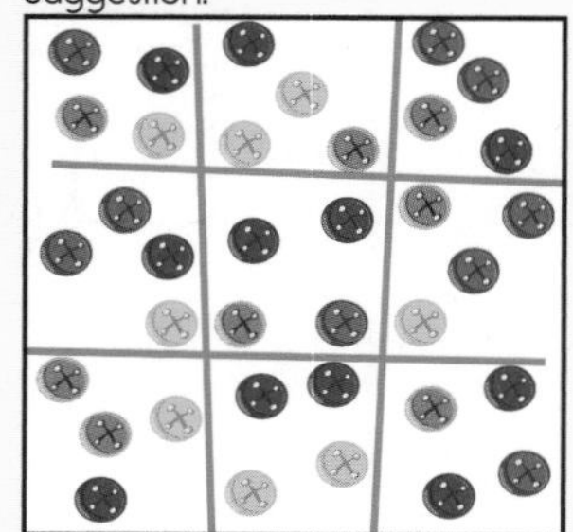

Page 49

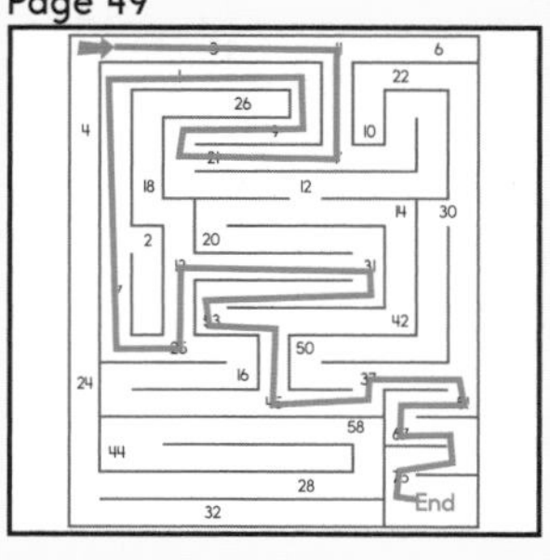

Page 50

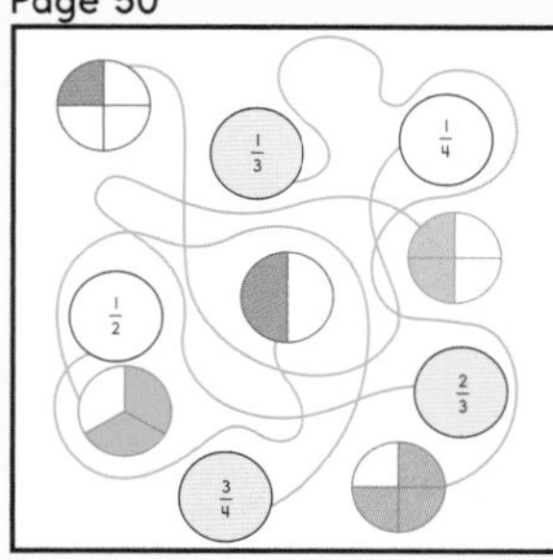

Page 51
Have someone check your answers.

Page 53

Page 56

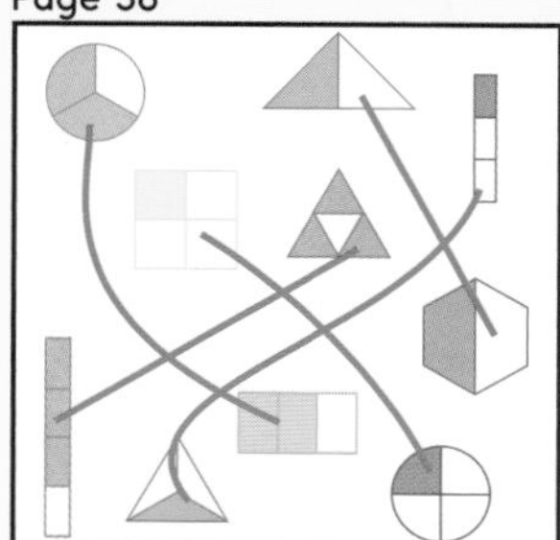

Page 57

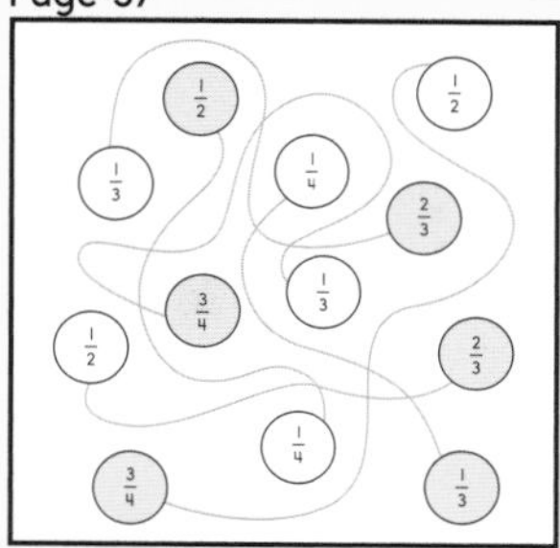

Answers

Page 58

1. $\frac{2}{3}$
2. $\frac{1}{3}$
3. $\frac{4}{4}$
4. $\frac{1}{2}$
5. $\frac{3}{4}$

Page 59

1. $\frac{1}{3}$
2. $\frac{3}{4}$
3. $\frac{2}{2}$
4. $\frac{1}{2}$
5. $\frac{2}{4}$
6. $\frac{3}{3}$
7. $\frac{2}{3}$
8. $\frac{4}{4}$

There was a WHOLE IN IT.

Page 60

1. PE
2. N
3. C
4. IL
5. E
6. R
7. AS
8. ER

PENCIL ERASER

Page 61

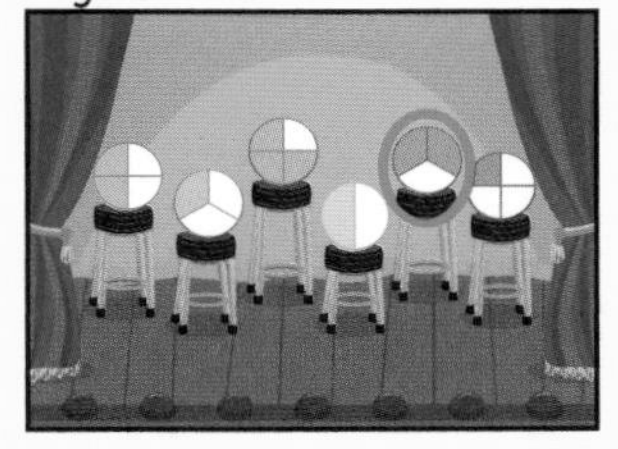

Page 62

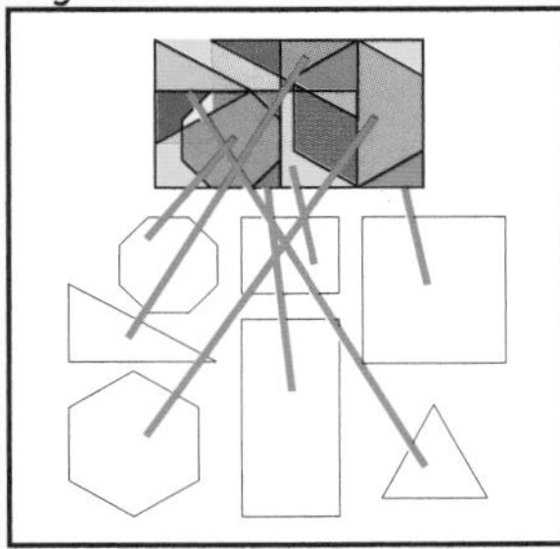

Page 64
Suggestion:

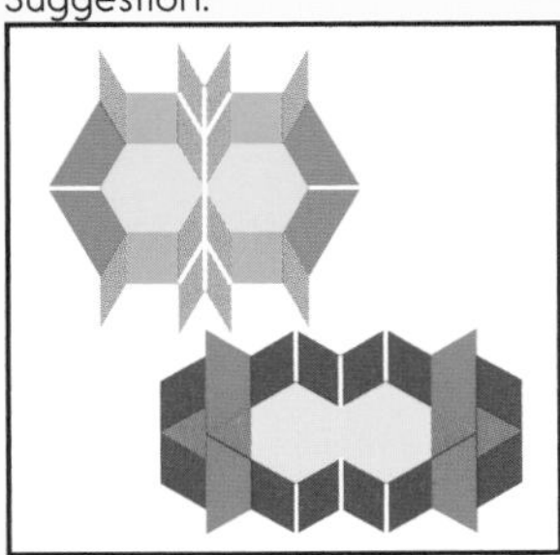

Page 65

Page 66
Suggestion:

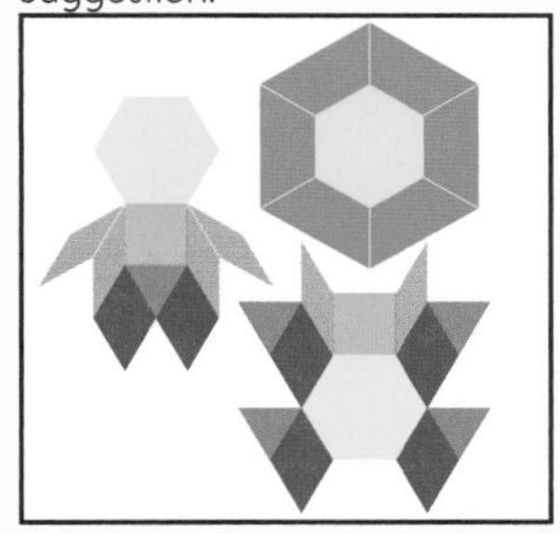

Page 67
Suggestion:

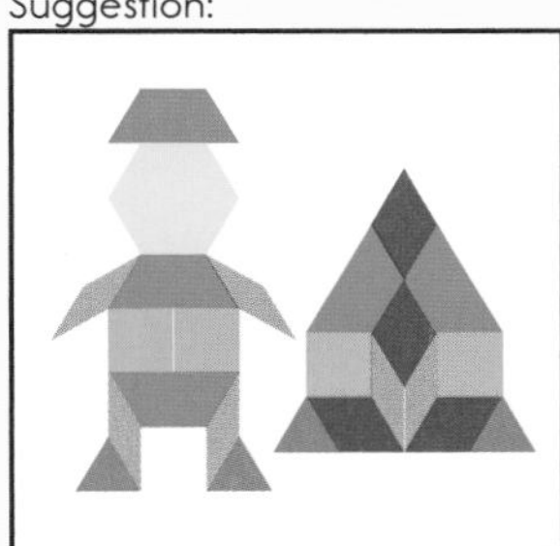

Page 68
Suggestion:

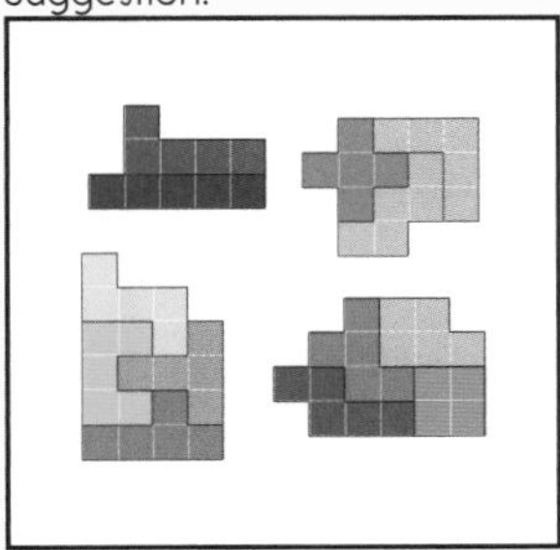

1. 16
2. 18
3. 20
4. 20

Page 69
Have someone check your answers.

Page 70
Suggestion:

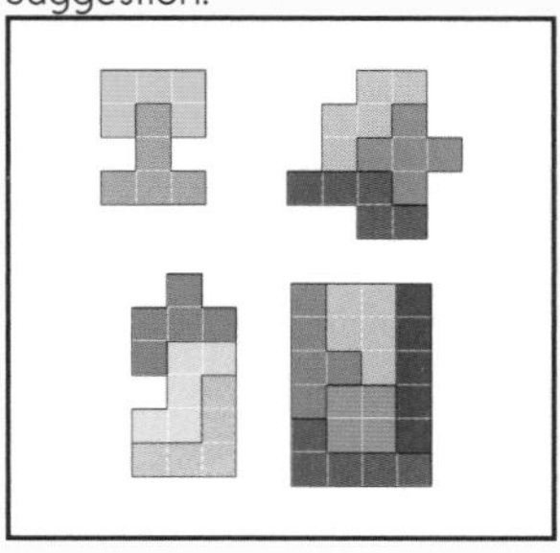

1. 10
2. 15
3. 15
4. 24

Page 71
Have someone check your answers.

Page 72

1. Monkeys
2. Penguins
3. Polar Bears
4. Elephants
5. Camels
6. Reptiles
7. Lions
8. Sea Lions
9. Zebras
10. Giraffes

Page 73

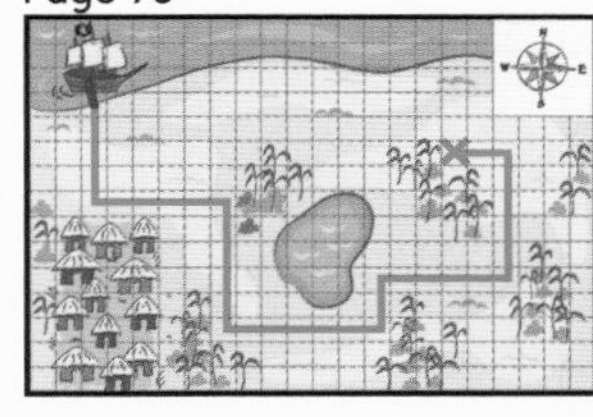

Page 74
Suggestion:

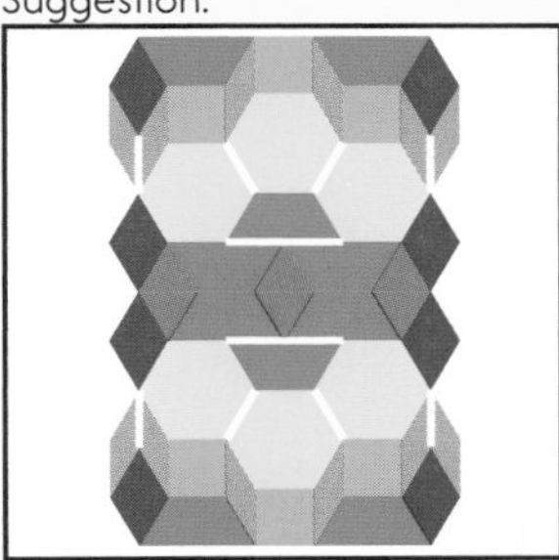

Page 75
Suggestion:

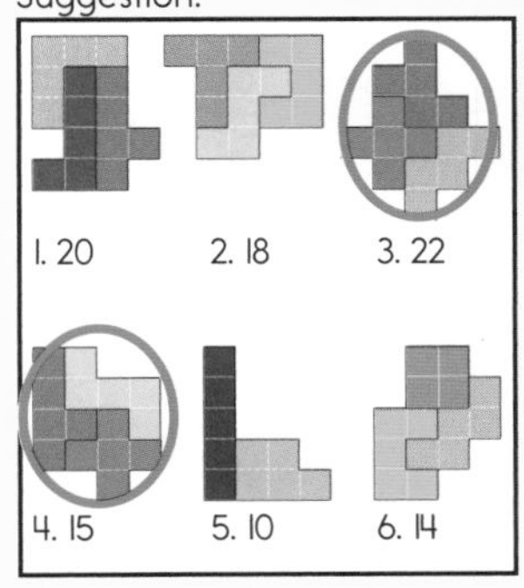

Page 76

1. quarter
2. nickel
3. dime
4. penny

Page 77

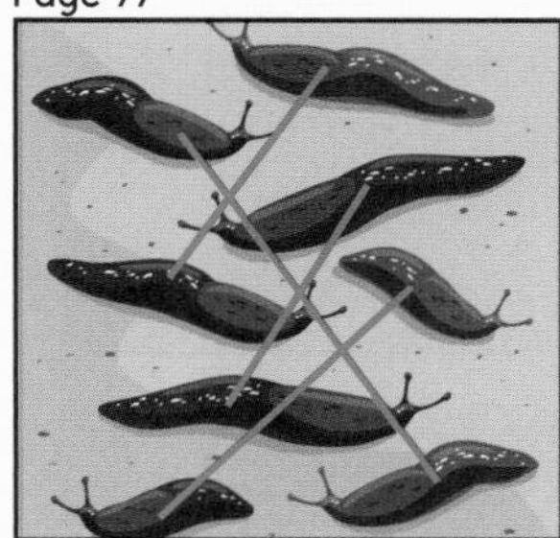

Page 78
A FOOTLONG.

Page 79

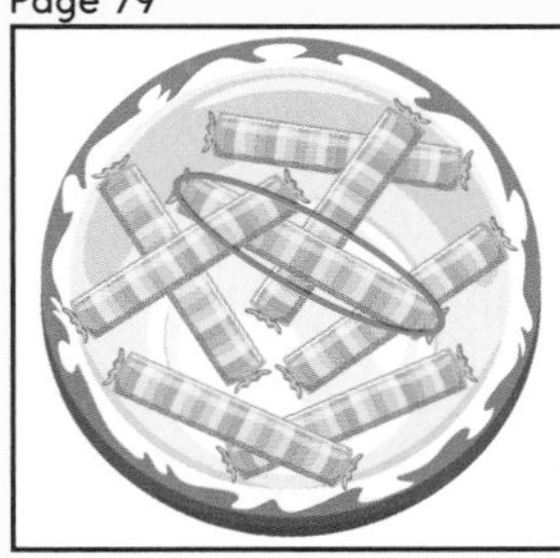

Page 80
IN SCENTIMETERS.

Page 81

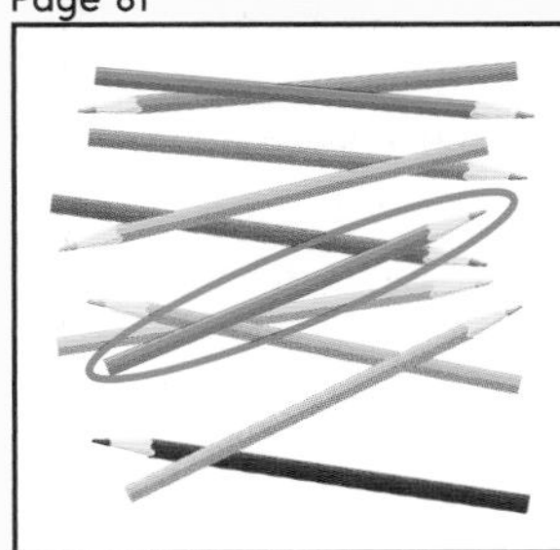

Page 82

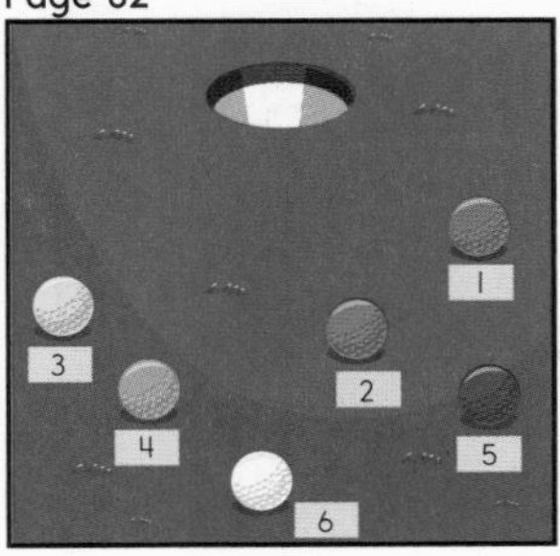

Page 83
Check:

1. 10
2. 12
3. 16
4. 13

Answers

Page 84

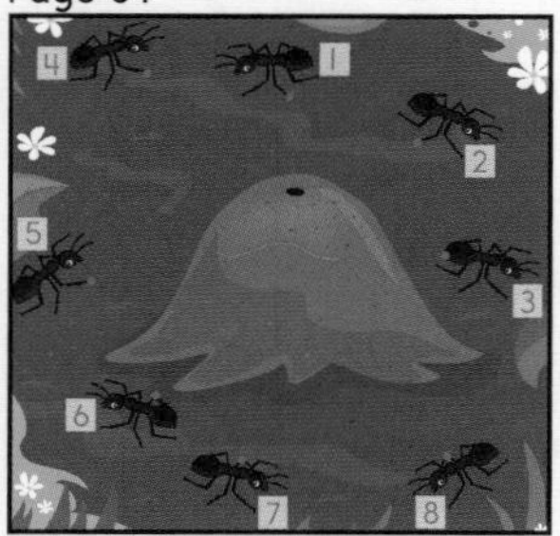

Page 85

Check:

1. 12	2. 10
3. 7	4. 8
5. 4	6. 13

Page 86

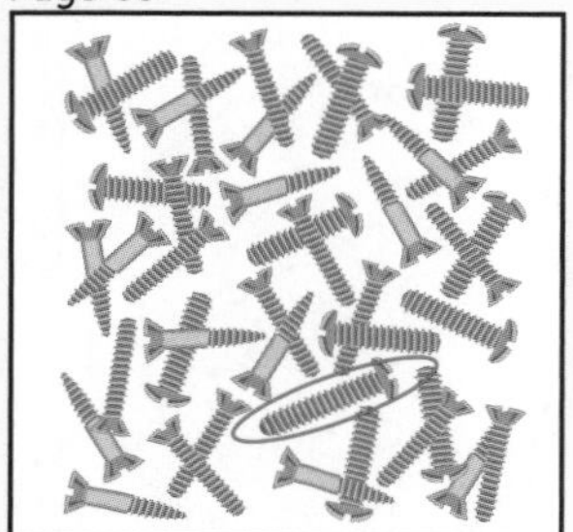

Page 87

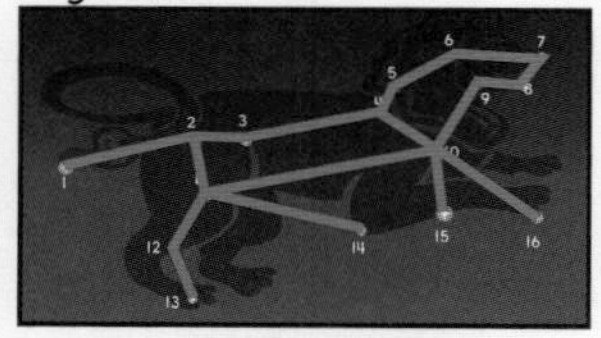

Check:

1. 2	2. 3
3. 8	4. 5
5. 10	6. 9

Page 88

A CLOCK.

Page 89

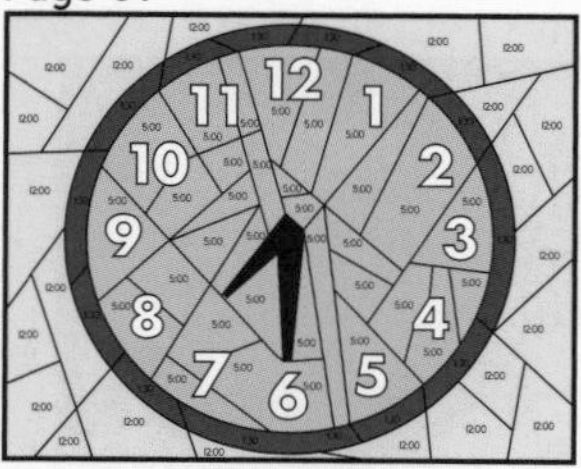

7:30

Page 90

THE LETTER E.

Page 91

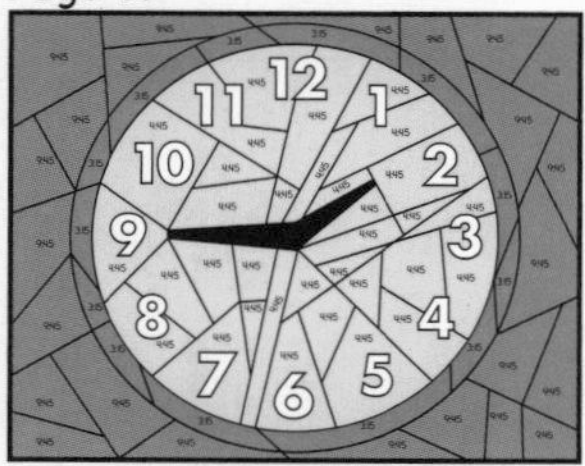

1:45

Page 92

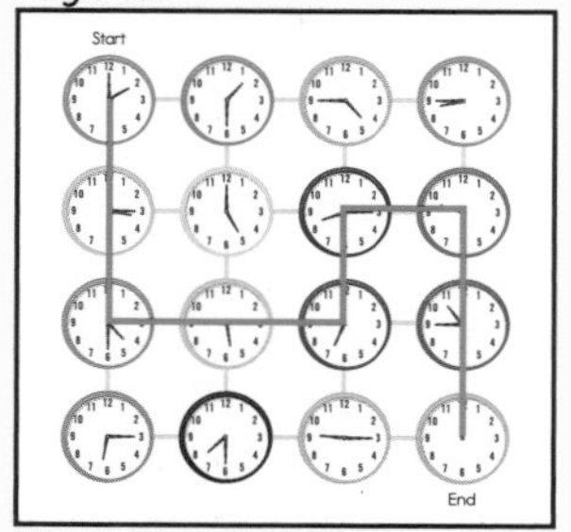

Page 93

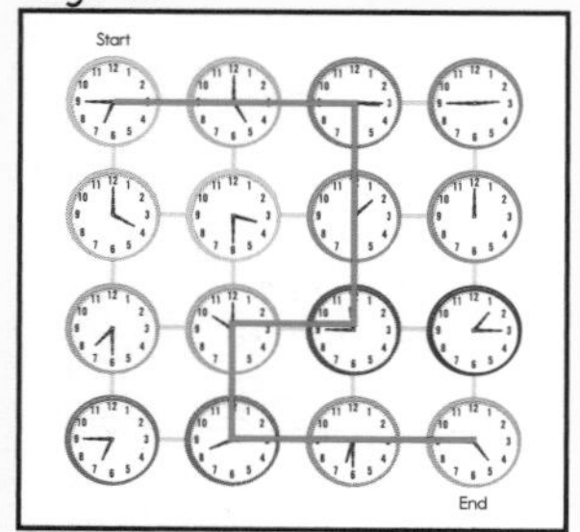

Pages 94–95

1. 1:00	2. 6:45
3. 5:15	4. 12:45
5. 8:00	6. 3:30
7. 2:30	8. 9:30
9. 11:15	10. 5:00
11. 2:15	

HE WANTED TO BE ON TIME.

Page 96

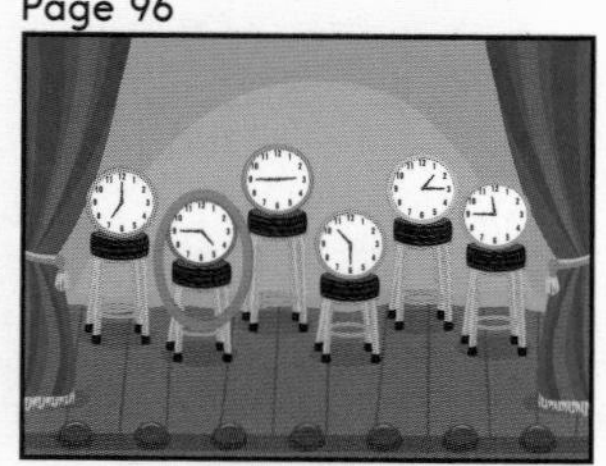

Page 97

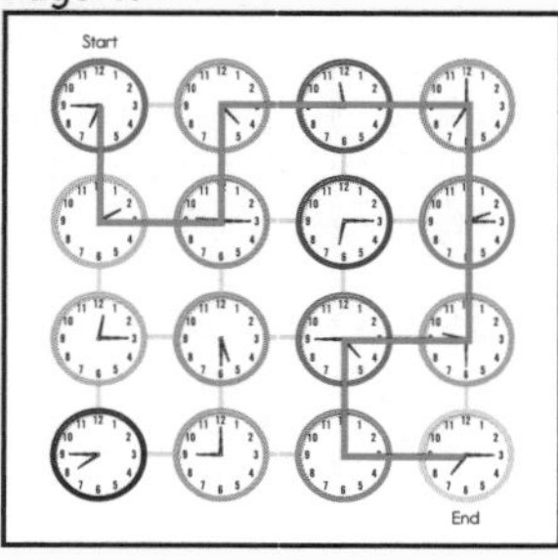

Page 98

1. 30¢	2. 22¢
3. 15¢	4. 51¢
5. 14¢	6. 75¢
7. 25¢	8. 40¢
9. 5¢	10. 41¢
11. 65¢	

IT HAS FOUR QUARTERS.

Page 99

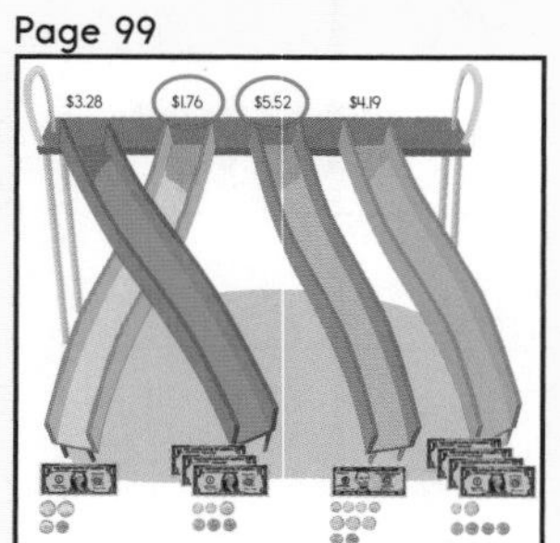

Page 100

1. $1.75	2. $2.02
3. $5.10	4. $6.15
5. $3.50	6. $1.23
7. $2.15	8. $5.60

You HAVE CHANGED.

Page 101

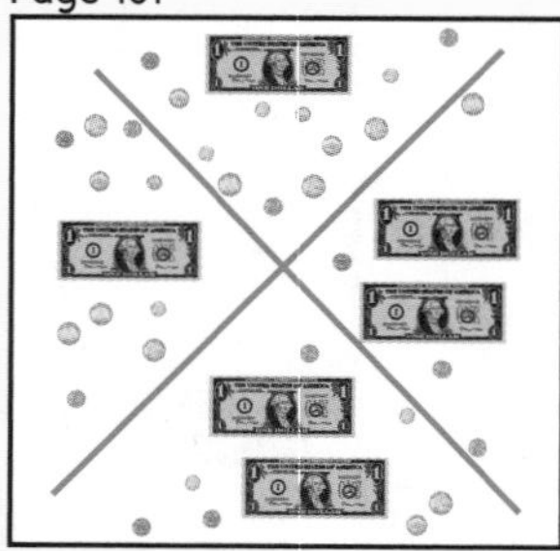

Page 102

Have someone check your answers.

Page 105

Extra: 

Page 106

1. $1.50	2. $3.07
3. $4.98	4. $6.19
5. $7.45	

A COIN.

Page 107

1. 4.98	2. 1.03
3. 6.37	4. 2.29
5. 5.14	

Pages 108–109

1. 1.41	2. 7.30
3. 3.13	4. 4.35
5. 1.21	6. 5.27
7. 3.75	8. 2.04
9. 1.60	10. 5.33
11. 6.45	12. 2.16
13. 1.26	14. 8.15

BECAUSE THE CHICKEN WAS ON VACATION.

Page 110